ECOLOGY

THE BASICS

The text sets out the fundamentals that underpin ecology – the study of the relations between organisms of the same of different species and between organisms and the abiotic (non-living) environment. In other words, it is the study of where we live.

The structure of the book adopts the classical style followed by most educators teaching this subject. It is divided into seven chapters which could be the basis of a semester's learning. Non-specialist, jargon-free language is used throughout with technical words and phrases explained in the text and in a glossary. Each chapter begins with an overview of the key themes covered and ends with suggestions for further, self-directed study. Chapter 1 is an introduction to the subject and contains outlines of how the subject has developed, its relevance today, and the skills required by an ecologist. Chapters 2–7 address, in turn, the abiotic (non-living) environment, adaptations of organisms to their abiotic environment, population ecology, community ecology, ecosystem ecology, and landscape ecology. Cross-cutting themes such as behavioural ecology, aquatic ecology, and terrestrial ecology are woven throughout these chapters. Relevant case studies, based on twenty-first century global challenges (changing climate, urbanisation, the biodiversity crisis, and the spread of invasive non-native species) and ecological ways of tackling these issues (nature-based solutions, and species and habitat conservation, including rewilding) are presented in relevant chapters.

In *Ecology: The Basics* the author synthesises ecological understanding to provide a foundation for first- and second-year undergraduates who are approaching ecology for the first time or are looking to refresh their knowledge. *Ecology: The Basics* also serves as a foundation for those coming to ecology as part of their higher-level studies, or who wish to be more informed about ecology as part of their life-long learning journey.

Philip James is Emeritus Professor of Ecology at the University of Salford, UK, and an Honorary Senior Research Fellow in Geography in the School of Education, Environment and Development, University of Manchester, UK. Prior to retiring, he taught ecology and related subjects for nearly 30 years. His research interests are in urban ecology, wildlife conservation, habitat management, and health and well-being linked to the natural environment. Since retiring he volunteers at two local country parks, is a trustee of the Lancashire Wildlife Trust, and plays the violin and sings with community orchestras and choirs.

The Basics

The Basics is a highly successful series of accessible guidebooks which provide an overview of the fundamental principles of a subject area in a jargon-free and undaunting format.

Intended for students approaching a subject for the first time, the books both introduce the essentials of a subject and provide an ideal springboard for further study. With over 50 titles spanning subjects from artificial intelligence (AI) to women's studies, *The Basics* are an ideal starting point for students seeking to understand a subject area.

Each text comes with recommendations for further study and gradually introduces the complexities and nuances within a subject.

POLYGAMY
BRUCE WHITEHOUSE AND WILLIAM JANKOWIAK

SUICIDE
MATIAS GAY

ROMANTICISM
ELIZABETH A. FAY

LEADERSHIP
MORGEN WITZEL

HUMAN-COMPUTER INTERACTION
ALAN DIX

ECOLOGY
PHILIP JAMES

For more information about this series, please visit: www.routledge.com/The-Basics/book-series/B

ECOLOGY

THE BASICS

Philip James

Routledge
Taylor & Francis Group
LONDON AND NEW YORK

Designed cover image: © Getty Images

First published 2027
by Routledge
4 Park Square, Milton Park, Abingdon, Oxon OX14 4RN

and by Routledge
605 Third Avenue, New York, NY 10158

Routledge is an imprint of the Taylor & Francis Group, an informa business

British Library Cataloguing-in-Publication Data
A catalogue record for this book is available from the British Library

Library of Congress Cataloging-in-Publication Data
Names: James, Philip (Professor of ecology) author
Title: Ecology : the basics / [Philip James].
Description: Abingdon, Oxon ; New York, NY : Routledge, 2026. |
Includes bibliographical references.
Identifiers: LCCN 2026023643 (print) | LCCN 2026023644 (ebook) |
ISBN 9781032994956 hardback | ISBN 9781032994970 paperback |
ISBN 9781003604464 ebook
Subjects: LCSH: Ecology
Classification: LCC QH541 .J36 2026 (print) | LCC QH541 (ebook)
LC record available at https://lccn.loc.gov/2026023643
LC ebook record available at https://lccn.loc.gov/2026023644

ISBN: 978-1-032-99495-6 (hbk)
ISBN: 978-1-032-99497-0 (pbk)
ISBN: 978-1-003-60446-4 (ebk)

DOI: 10.4324/9781003604464

Typeset in Bembo
by SPi Technologies India Pvt Ltd (Straive)

CONTENTS

ACKNOWLEDGEMENTS

As I was pulling together my initial thoughts and preparing the proposal for this book, I sought advice and guidance from Dr Kostantinos (Kosta) Tzoulas. Kosta had been a student of mine before taking up a post as a Senior Lecturer at Manchester Metropolitan University, UK. The adage that 'the teacher learns from their student' was never more apt. Sadly Kosta, aged 47, died in the spring of 2025.

As the writing progressed, I sought advice and guidance ranging from structural issues, a specific point, or the depth of the treatment I was proposing giving to a particular topic from Dr Matthew Dennis, another former PhD student who is currently Reader in Spatial Ecology at the University of Manchester, UK.

Matthew was one of three people who read an early draft of this book. His comments, along with those of Martin Brown and Alison Elizebeth Larkin, helped iron out many issues relating to the content and the way I had expressed my ideas.

Following a process of iterative re-drafting, the text was read again, this time by Professor Sheila Pankhurst (University of Cumbria, UK), Professor Rob Young (University of Salford, UK), and Professor Jonny Huck (University of Manchester, UK). Their comments prompted me to delete text that was superfluous to requirements, to plug gaps where they were perceived, and to improve the clarity of the text where that was needed.

Upon submission of the complete copy of the text to the publisher, it was sent out for review by two anonymous ecologists working in higher education. Both reviewers provided insightful and helpful comment. I have endeavoured to address all their comments while maintaining the breadth of the topics covered in the book.

When it came to time to find photographs to illustrate points within the text, I turned to colleagues and friends: Barbara Firth, the estate of Brain Rogers (arranged via Rob Young), Carole Conroy, Charlotte Weinberg, David James, Elmeri James Vainikainen Dennis, Gina Hazell, Jan Sayle, Luke Blazejewski, Matthew Dennis, Mike Allen, Rob Young, and Sue Melville. Their generosity in sharing these photographs elevated the contents of the book immeasurably.

GLOSSARY

Abiotic factors the non-living components of an ecosystem that influence the environment and the organisms within it.

Abundance see *species abundance*.

Aestivation (or estivation) a state of animal dormancy that occurs in response to high temperatures and arid conditions (cf. *hibernation*).

Apex predators consumers that are at the top of the food chain and have no natural predators.

Autotrophs see *producers*.

Behaviour how organisms interact with their environment and each other through movement and social interactions (e.g. communication, courtship, raising young, and dominance).

Biodiversity the variability among living organisms from all sources including, *inter alia*, terrestrial, marine and other aquatic ecosystems and the ecological complexes of which they are part; this includes diversity within species, between species, and of ecosystems.

Biodiversity hotspot a biogeographic region that is both rich in endemic species (i.e. species that are found nowhere else) and significantly threatened by human activities (e.g. agriculture, natural resource exploitations, or urbanisation).

Biogeographical realms (regions) the broadest biogeographical division of the Earth's land surface, based on plant and animal distributions having similar or shared characteristics.

Biomes ecological communities that have the same climatic conditions and geologic features and that support species with similar life strategies and adaptations. Biomes are the fundamental units of biogeographical realms/regions.

Biotic factors the living components (such as plants, animals, fungi, and bacteria) of an ecosystem. They play critical roles in ecosystems as producers, consumers, and decomposers, contributing to the energy flow and nutrient cycling necessary for life.

Carrying capacity (K) the maximum number of individuals that an area can support.

Chemosynthesis the biological process by which certain micro-organisms, in the absence of light, convert inorganic carbon-containing molecules (usually carbon dioxide (CO_2) or methane (CH_4)) and nutrients into organic matter using the oxidation of inorganic compounds (e.g. hydrogen gas (H_2), hydrogen sulphide H_2S)) or ferrous ions (iron III ions) as a source of energy.

Climate the general weather conditions (e.g. temperature, precipitation (type and amount), humidity, hours of sunshine, cloud cover, atmospheric pressure, fogginess, and wind velocity) usually considered over a period of at least ten years, found in a particular place (cf. *weather*).

Climate change a change in global or regional climate patterns.

Community a group or association of populations of two or more different species occupying the same geographical area at the same time.

Competition interactions between organisms of the same (intraspecific competition) or different (interspecies competition) when there is insufficient of a given resource to meet the demands of all the organisms.

Consumers organisms that are unable to make their own energy and rely on the consumption and digestion of producers or other consumers, or both, to survive. Primary consumers are herbivores that consume primary producers, secondary consumers are carnivores that eat primary consumers, and tertiary consumers are higher-level carnivores that eat secondary consumers.

Coriolis effect the apparent deflection of moving objects not firmly connected to the ground caused by the rotation of the Earth. Although the earth rotates once per day, the speed at which the earth's surface moves in is greater at the equator (around 1,600 km hr^{-1}) than it is at the poles around 0.00008 km hr^{-1}). Air and water are not firmly connected to the ground and as the earth rotates, they tend to be deflected due to the movement of the Earth. This apparent deflection influences weather patterns and ocean currents.

Decomposers organisms that break down dead organic material converting it into simpler inorganic materials, thus, making nutrients available to primary producers. They include bacteria, fungi, and detritivores (see *Detritivores*).

Deductive the inference of particular instances from general laws (cf. *inductive*).

Dependent variable the variable (factor that can change) being tested in a scientific experiment, the variable that the research is interests in understanding, predicting, or explaining based on the changes made to the independent variable) that is manipulated of controlled in a scientific experiment (cf. *independent variable*).

Detritivores organisms that obtain nutrients by consuming detritus (i.e. decomposing plant and animal matter as well as faeces). Detritivores play a crucial role in ecosystems by recycling nutrients and are found at various trophic levels, feeding on material from primary producers, herbivores, and carnivores. Common examples of detritivores include earthworms (Annelida), springtails (Collembola), millipeds (Diplopoda), and certain crustaceans (Crustacea).

Dicotyledons (aka **dicots**) one of the two major groups of flowering plants (the other being monocotyledon) characterised by having two cotyledons (embryonic leaves) in their seeds. Typically, dicotyledons have floral organs (e.g. anthers and petals) arranges in cycles of four or five and have leaves with a reticulate (net-like) venation pattern (cf. *monocotyledons*).

Ecological trap scenarios in which rapid environmental change leads organisms to prefer to settle in poor-quality habitats. Ecological traps occur when the attractiveness of a habitat increases disproportionately in relation to its value for survival and reproduction.

Ecology a combination of two Ancient Greek words Oikos meaning 'the building, land and people which go together and the management thereof' and logos meaning 'study of'.

Ecosystem a biological community of interacting organisms and their physical environment.

Ecotone a region of transition between two biological communities.

Environment the air, water, and land in or on which organisms live.

Environmental management the management, design, and execution of processes designed to conserve, protect and manage resources and ecosystems sustainably. It includes policy-making and techniques to reduce pollution, address climate change, and protect biodiversity.

Evenness see *species evenness*.

Exponential an exponential rate of increase is one that becomes quicker and quicker as the thing that increases becomes larger. This can be compared with straight line growth in which the rate of increase remains the same.

Fecundity the ability to produce offspring or new growth.

Fragmentation (habitat fragmentation) occurs when parts of a habitat are destroyed, leaving behind smaller unconnected areas.

Gametophyte one of the two alternating multicellular phases in the life cycles of plants and algae. It is the sexual phase.

Geographic Information Systems (GIS) a mapping technology used to capture, store, manage, and analyse various types of geographical and ecological data.

Hibernation a state of animal dormancy that occurs in response to low temperatures (cf. *aestivation*).

Homeotherm an animal that maintains its body temperature at a constant level, usually above that of the environment, by its metabolic activity.

Hypothesis a testable statement based on a supposition or proposed explanation for a phenomenon made on the basis of limited evidence as a starting point for further investigation.

Independent variable the variable (factor that can change) that is manipulated of controlled in a scientific experiment (cf. *dependent variable*).

Inductive the inference of general laws from particular instances (cf. *deductive*).

Inorganic not consisting of or deriving from living matter. Compounds which are not organic (broadly, compounds not containing carbon combined with hydrogen and sometimes also with other elements such as oxygen and nitrogen).

Interbirth interval (IBI) the time between giving birth and a female becoming pregnant again (aka pregnancy spacing or inter-pregnancy interval).

Invasive non-native species (INNS) plants and animals which have been introduced (accidentally or deliberately) outside of their native range through human activity and which spread rapidly and cause damage to the environment, economy or human health.

Island biogeography examines the factors that affect the species richness and diversification of isolated natural communities. Originally developed to explain the pattern of the species–area relationship occurring in oceanic islands, the concept has been applied to terrestrial areas.

K**-selection species** a concept that describes a reproductive strategy used by certain species. It is part of the *r*/*K* selection theory, which categorises species based on their reproductive traits and survival strategies. Key Characteristics of *K*-selected species are long lifespan, adapted to stable environments, low reproductive rate, provision of parental care, and low mortality rate.

Landscape all the features of an area of land.

Metapopulation a group of spatially separated populations of the same species that interact at varying levels. Metapopulation theory is a framework for understanding the dynamics of species that inhabit fragmented landscapes.

Methodology the techniques, tools, and procedures used to collect and analyse data or information.

Minerals a solid, naturally occurring inorganic substance.

Model (ecological model) a representation of an ecological system that is studied to better understand the real system. Usually, a mathematical or abstract form of a highly complex system.

Monocotyledons (**aka monocots**) one of the two major groups of flowering plants (the other being dicotyledon) characterised by having one cotyledon (embryonic leaf) in their seeds. Typically, dicotyledons have floral organs (e.g. anthers and petals) arranges in cycles of three and have leaves with a parallel venation pattern (cf. *dicotyledons*).

Morphology the branch of biology that deals with the form of living organisms.

Mortality the ratio of the number of deaths to the size of a population.

Mycorrhiza a (generally) mutually beneficial relationship between the root of a plant and a fungus that colonises the plant root.

Natality the ratio of the number of births to the size of the population: birth rate.

Natural history the study of organisms in their natural environment, focusing on observational methods to understand the diversity and interconnectedness of life. It has a is a strong multidisciplinary nature incorporating many branches of sciences including geology and ecology.

Nature-based solutions (NbS) the use of nature (biodiversity) and natural processes to address diverse socio-environmental issues including climate change mitigation and adaptation, human security issues such as water security and food security, and disaster risk reduction.

Net primary productivity the difference between photosynthesis (gross primary production) and respirations (energy usage) in an ecosystem.

Niche the role and position a species has in its environment, encompassing how it interacts with biotic (living) and abiotic (non-living) factors. It describes how a species survives, reproduces, and utilises resources within its habitat, including its interactions with other species, such as competition and predation.

Niche portioning the concept that competing species have different patterns of resource use or different niches and, thereby, avoid competing with one another.

Nutrients a substance (e.g. carbohydrates, lipids, proteins, vitamins, and minerals) that provides nourishment essential for the maintenance of life and for growth.

Organic matter carbon-based compounds containing hydrogen and, in many cases, other elements (e.g. oxygen and nitrogen) found in nature including carbohydrates, lipids, proteins, and nucleic acids.

Patch–corridor–matrix model describes the horizontal landscape structure. The patch comprises the individual landscape elements embedded in the matrix (e.g. hedges, settlements, woods). The corridor is the linear, connecting elements (e.g. river network). The matrix is the predominant type of land use (e.g. grassland) within which patches and corridors are set.

pH a logarithmic scale used to specify the acidity or basicity of aqueous solutions. Hence, to measure the pH of dry soil or tree bark it is necessary to add water.

Photoperiodism the functional or behavioural response of an organisms to changes of duration in daily, seasonal, or yearly cycles of light and dark.

Photosynthesis the biological process by which green plants, algae, and some bacteria convert light energy into chemical energy.

Phylogenetic the scientific study of evolutionary relationships among biological entities, often species, individuals, or genes. A phylogenetic tree is a graphical representation illustrating how species have evolved from common ancestors.

Physiology the branch of biology that deals with the normal functions of living organism and their parts (e.g. organs).

Poikilotherm an animal that cannot regulate its body temperature except by behavioural means such as basking or burrowing (cf. *homeotherm*).

Population a subset of individuals of one species that occupies a particular geographic area and, in sexually reproducing species, interbreeds.

Producers also known as autotrophs or primary producers, are organisms that create their own food through photosynthesis or chemosynthesis. Producers are the foundation of ecosystems and food webs.

r**-selection species** a concept that describes a reproductive strategy used by certain species. It is part of the *r*/*K* selection theory, which categorises species based on their reproductive traits and survival strategies. Key characteristics of *r*-selected species are short lifespan, adaptation to unstable environments, high reproductive rate, minimal parental care, and high mortality rate.

Rescue effect the migration of individuals such that small, isolated populations persist, metapopulations are stabilised, and the chance of extinction is reduced.

Richness see *species richness*.

Salinity the saltiness or amount of salt dissolved in a body of water.

Seasonality the variation in phenomena depending on the season or time of year.

Sere a sequence of ecological communities that develop over time during ecological succession.

Shrub a woody plant which is smaller than a tree and has several main stems arising at or near the ground (cf. *tree*).

Stratification refers to the vertical layering of a habitat (e.g. the vertical layers within a forest or on a rocky shore).

Social-ecological a framework that draws from natural ecosystems to understand the dynamic interrelations between and interdependence of individuals and groups, and their environment.

Species abundance the total number of individuals from a given species within a specific area (e.g. a community, ecosystem, or landscape).

Species evenness the relative abundance of different species within a specific area (e.g. a community, ecosystem, or landscape).

Species richness the number of species within a specific area (e.g. a community, ecosystem, or landscape).

Succession (**aka ecological succession**) the gradual process by which communities and ecosystems change over time. Primary succession occurs in lifeless areas where no soil exists (e.g. after a volcanic eruption, following glacial retreat, or the laying of tarmac, or concrete). Secondary succession occurs where a disturbance (natural or human induced) has destroyed an existing community but left the soil intact (e.g. after a flood or fire, or following deforestation or other land clearance activities).

Taxon a taxonomic group of any rank (e.g. species, genus, family, class, order).

Thermocline the transition layer between warmer water at the top of a water body (a lake or pond) and the cooler deep water below.

Topography the arrangement of the natural and artificial physical features of an area (some examples include rivers, mountains, cliffs, valleys, skyscrapers).

Torpor a state of physical or mental inactivity.

Transpiration the loss of water, mainly through stomata, from a plant.

Tree a woody perennial plant, typically having a single stem or trunk growing to a considerable height and bearing lateral branches at some distance from the ground (cf. *shrub*).

Trophic relating to feeding and nutrition.

Trophic levels refer to the hierarchical positions of organisms in a food chain or food web, based on their feeding relationships.

Weather the conditions (e.g. temperature, precipitation (type and amount), humidity, hours of sunshine, cloud cover, atmospheric pressure, fogginess, and wind velocity) at a particular time in a particular area (cf. *climate*).

1

ECOLOGICAL KNOWLEDGE AND SKILLS AND THEIR APPLICATION

1.1 INTRODUCTION

Ecology is a relatively young science: the word 'ecology' was first used in 1866. Since then, numerous sub-disciplines have emerged many of which are listed in the first section of this chapter. What ties the sub-disciplines together is a focus on various aspects of the relations of organisms to one another and to their physical surroundings.

The purpose of this book is to provide a foundation for everyone interested in the natural world. Each chapter is designed to be 'stand alone'. While a chapter may be read in isolation, there is much to be gained by following the book from cover to cover as that allows the reader to integrate ideas from the different topics into a broader understanding of ecology. Those interested in building on these foundations can then go on to acquire specialist knowledge according to their personal interests.

In each chapter relevant ecological principles, theories, and models are set out. Chapter 1 is an overview of the development of ecological thought and the emergence of a myriad of subdisciplines, the skills required by an ecologist, and the role ecologists play in policy development. The remaining chapters follows a hierarchy starting with the physical environment (Chapter 2), moving on to individual organism (Chapter 3), populations (Chapter 4), communities (Chapter 5), ecosystems (Chapter 6), and landscapes (Chapter 7). This means that topics such as behavioural ecology are interwoven through the book (e.g. behaviours that are between members of the same species are addressed in Chapter 4 (populations), while

DOI: 10.4324/9781003604464-1

behaviours that occur between different species are addressed in Chapter 5 (communities)).

Alone and as part of interdisciplinary teams, ecologists tackle key societal issues including the changing climate, urbanisation, the global loss of biodiversity, and the spread of invasive non-native species (INNS). Ecologists also provide solutions to address these challenges including nature-based solutions, and species and habitat conservation programmes. For example, ecological knowledge and insights are essential for preserving species and habitats and maintaining functioning ecosystems which provide many of the basics required for a healthy life (including, clean air to breathe, clean water to drink, and food to eat). Devising an appropriate conservation strategy may involve examining, altering, and monitoring the physical environment that the species of interest lives in, controlling pests and diseases, recording and monitoring the species' genetic diversity, monitor population growth and community change, and supporting local communities who live and interact with the focal species. The last section in each chapter is a case study highlighting how the knowledge presented in that chapter is related to one of the global challenges and to actions to address that challenge.

In the second section of this chapter the skills required by ecologists are set out. Ecologists typically possess a range of valued skills in addition to the academic knowledge garnered through study. There are both, so called, hard and soft skills. Hard skills (e.g. survey techniques, DNA profiling, and species identification) are learnt and acquired through practical work and experience, alongside academic study. The soft skills (e.g. communication and interpersonal skills) come to the fore when ecologists are working in interdisciplinary teams and seeking to communicate their insights and influence the policy development and the implementation of ecologically based interventions. Increasingly, ecologists are part of an interdisciplinary team. The different ways that people from varied discipline interact with each other in these teams are explored in the third section of this chapter.

Many ecologists are not just experts on nature they also advocate for it. To do so requires an understanding of some aspects of human nature, psychology, and politics and of how human societies and individuals make decisions. The final section in this chapter comprises a case study examining the relationship between ecological-based knowledge and policy. Barriers to the adoption of ecological knowledge into policy are identified and a means to overcome these, though better and different communication strategies, is outlined. This aids in understanding how ecologists work alongside those from other disciplines to bring about policy change and associated actions.

1.2 ECOLOGY – ORIGINS AND SUB-DISCIPLINES

In 1866, Ernst Haeckel (1834–1919, a German zoologist, naturalist, philosopher, physician, marine biologist, and artist) officially coined the term 'ecology' (from the Greek *oikos*, meaning 'household', and *logos*, meaning 'study'; Haeckel 1866, II: 286–289). This formalisation marked the birth of ecology as a distinct scientific field.

The goal of ecology, today, is to understand the patterns and processes related to life on earth – how plants and animals have adapted and continue to adapt to changing environments and to the presence of new species. Ecologists initially addressed this goal using an observational and descriptive approach. Then, in the 1970s there was a marked move towards an approach that is experimental, empirical, and predictive. Adopting this newer approach, ecologists formulate possible explanations for the phenomena they observe, frame these explanations as hypotheses, (i.e. as testable statements) which they test by experimentation in the laboratory or the field. From their results ecologists look for patterns in their data, make models and predictions, and draw generalities. One such set of generalities, the four laws of ecology, were formulated in 1971 by Barry Commoner (1917–2012, an American cellular biologist, college professor, and politician and one of the founders of the modern environmental movement). These four laws are a set of rules that govern natural systems and are a foundational framework for understanding the nature of ecological interactions and the consequences of human activity and laid the groundwork for modern ecology:

1. Everything is connected to everything else – hence, a change in one part of an ecosystem ripples throughout the whole system,
2. Everything must go somewhere – matter and energy cannot be destroyed and will accumulate or cycle through different forms,

3. Nature knows best – human interventions in ecosystems, which have developed over millions of years, often backfire, and
4. There is no such thing as a free lunch – every resource exploited by people carries an ecological cost that must be paid somewhere, sometime.

Since 1866, ecology has broadened in its scope and many sub-disciplines have emerged named according to factors such as spatial scale, taxon, biome, and methods (Table 1.1). Sub-disciplines are not mutually exclusive and there is often overlap between two of more areas of study. For example, terrestrial ecology includes all aspects of ecology related to the terrestrial environment including population ecology. At the same time, population ecology covers all aspects of populations across all environments including the terrestrial environment.

Table 1.1 Examples of sub-disciplines of ecology

Division of ecology according to	*Examples of sub-disciplines*
Biogeographical realm or climate area	Arctic ecology, polar ecology, tropical ecology
Biological classification or taxon	Animal ecology, human ecology, insect ecology, microbial ecology, plant ecology
Biome	Aquatic ecology, desert ecology, estuarine ecology, freshwater ecology, grassland ecology, marine ecology, urban ecology
Ecological aspects or phenomena	Behavioural ecology, chemical ecology, disease ecology, ecophysiology, ecotoxicology, evolutionary ecology, fire ecology, functional ecology, genetic ecology, landscape ecology, limnology, molecular ecology, paleoecology, sensory ecology, social ecology, soil ecology, spatial ecology, and thermal ecology
Level of organisation or scope	Autecology, community ecology, ecosystem ecology, population ecology, synecology
Methodology used for investigation	Field ecology, theoretical ecology
Spatial scale of the ecological system	Global ecology, macroecology, microecology

It can be seen from the information in Table 1.1 that ecologists study aspects of the physical and biological environment. The physical environment encompasses all abiotic ('abiotic' comes from the Greek and means 'non-living) elements such as landforms, climate, water bodies, soil, and infrastructure, as well as the natural processes that shape them, including erosion, earthquakes, and volcanic activity. The physical environment affects the physiology, growth, reproduction, and survival rates of species. Ecologists investigate and monitor the physical environment (see Chapter 2) and how species are adapted to their environment and are adapting to changes in that environment (see Chapter 3).

The biological or biotic environment ('biotic' comes from the Greek meaning living) includes all living organisms. Ecologists study population and community dynamics (e.g. changes in species size, density, and distribution), relationships within populations and communities (e.g. competition, mutualism, and predation) and how environmental disturbances influence population growth or decline and community composition (Chapters 4 and 5). When the abiotic and biotic factors of the environment are combined into a functional unit, that unit is called an ecosystem. Energy flows through an ecosystem from the sun through producers (e.g. green plants) which convert solar energy into chemical energy (in the form of glucose. From there the chemical energy flows on through herbivores and carnivores to decomposers (which break down organic matter releasing the stored energy). Materials (e.g. essential nutrients, such as carbon, nitrogen, and phosphorus) cycle around an ecosystem between living organisms and the environment in processes including the carbon and nitrogen cycles. Ecologists study these flows and movements and the services and benefits that derive from them (Chapter 6). Ecosystems exist alongside each other and form a landscape. Ecologists study the flows of species, energy, and materials between different ecosystems and how these landscapes function (see Chapter 7). Today, ecologists from all sub-disciplines are involved in addressing global issues such as the loss of biodiversity, climate change, habitat loss, and the emergence and spread of pests and diseases affecting plants and animals (including humans). Ecological knowledge and understanding are important in informing policy and practical actions aimed at tackling these issues.

1.3 ECOLOGICAL KNOWLEDGE AND SKILLS

A list of the common personal characteristics shared by ecologists are presented in Table 1.2.

This list illustrates how knowledge comes together with practical skills and experiences, and personal traits in a professional ecologist (Figure 1.1). Anyone wishing to pursue a career in a particular sub-discipline will need to acquire specialist knowledge and skills. It is also important for an ecologist to obtain experience which may be gained through volunteering, employment, and/or training courses to build up specific understanding of the habitat(s), species, and/or ecological techniques.

1.3.1 SOFT AND HARD ECOLOGICAL SKILLS

Ecological skills can be divided broadly into soft and hard skills. Soft skills relate to personal attributes and include, but are not limited to, how an individual interacts with others, solves problems, and manages work. Soft skills are often underrated, but they can have a profound impact on performance. Hard skills refer to job-specific technical expertise, covering anything from species identification to data modelling. Together, both types of skills make up the professional toolkit of an ecologist.

Table 1.2 Common personal characteristics shared by ecologists

Personal characteristics
Interest in the natural world
Knowledge of the functioning of natural systems
Good academic qualifications in biological or environmental subjects
Expertise in one or more groups of living organisms
Enjoyment of fieldwork
Staying power to complete tedious and sometimes uncomfortable tasks in the field or laboratory
Ability to take an objective approach to conservation issues

Source: based on the Chartered Institute for Ecology and Environmental Management (CIEEM), an international professional membership body representing and supporting ecologists and environmental managers based in the UK (https://cieem.net)

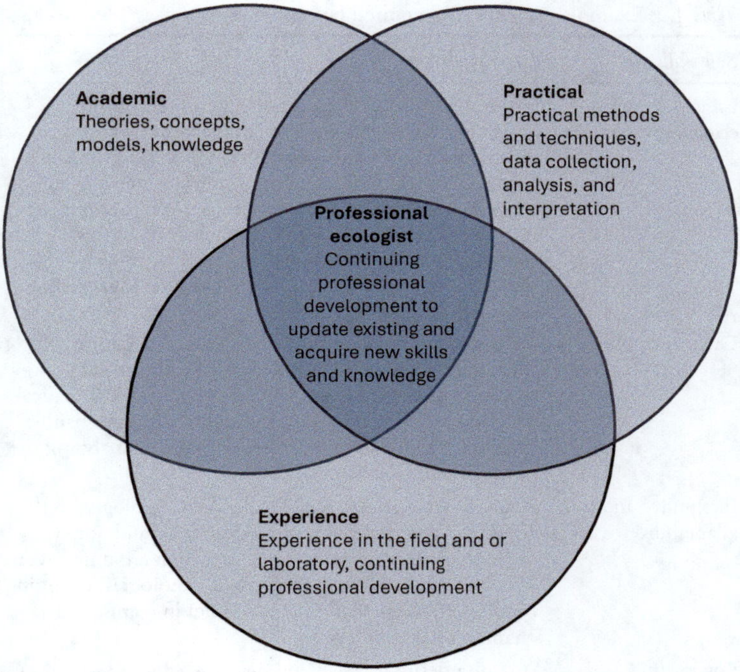

Figure 1.1 In their working life, a professional ecologist combines three domains: academic, practical, and experience.

Soft skills empower an ecologist, especially when they are in collaborative and public-facing roles and help facilitate the acquisition and development of hard skills. Some of the most sought-after soft skills are set out in Table 1.3.

While most ecologists will pass their careers encountering little or no conflict, this is not the always the case. Conflict can arise between individuals or groups with different cultures, vested interests, and views on, for example, the re-introduction of a species, habitat restoration techniques. Conflict can also occur in relation to access to a given area. When individuals from at-risk backgrounds (individuals from minority identities including but not limited to

Table 1.3 Examples of soft skills required by ecologists

Soft skills	*Description*
Effective communication	Communication is the cornerstone of all soft skills. Communication can be in writing, spoken, via illustrations, and increasingly through various social-media channels. Effective communication facilitates expressing complex ecological concepts in a way that is understandable for various audiences including individuals and groups who have little or no prior understanding of the issue, as well as individuals with extensive knowledge.
Critical thinking	Critical thinking is the ability to assess a situation thoroughly, evaluate available options, and make a reasoned judgment. For example, ecologists use critical thinking when making decisions about how to handle a sudden change in an ecosystem or when analysing the potential effects of a new policy.
Flexibility and adaptability	Being flexible and adaptable allows the ecologist to adjust to changing field, policy, funding, and /or technological conditions many of which are unforeseen. For example, incorporating new technologies, changing funding policies, changing personnel in teams, and changing stakeholders.
Curiosity	A strong desire to learn and develop knowledge and understanding. To ask questions and seek answers. To challenge existing knowledge.
A real passion for the natural world	An intense enthusiasm for one or more aspects of the natural world. For example, of a species, a group of species, a habitat (e.g. marine or woodland habitat), or for education or habitat management.

race/ethnicity, sexual orientation, disability, gender identity, and/or religion) enter unfamiliar communities to do fieldwork, they may be placed in an uncomfortable and potentially unsafe 'othered' position, and prejudice may manifest against them (Beltran et al. 2020). Prejudice-driven conflict can threaten a researcher's immediate and long-term physical health and safety with implications for their mental health, productivity, and professional development (Demery and Pipkin 2021). Demery and Pipkin (2021) provide details of strategies for researchers, supervisors, and institutions to minimise risk.

Soft skills are also important in the changing field of conservation management. Delivery of ecological conservation programmes has seen a move away from them being implemented 'for', to implementation 'with' and, finally, 'by' the local community. Conservation 'for' the local community occurs when people and groups from outside that community design and implement a conservation programme with little or no involvement of the local community. Conservation 'with' the community involves effective consultation programmes which empower the local community to work alongside those from outside to design and implement the programme. The final step is when a local community leads the design and implementation of the programme with external support and facilitation to acquire relevant skills and knowledge provided only when requested by the local community. This may require an ecologist to establish relevant education and upskilling programmes and to identify people and organisations who can provide external guidance as requested by the community.

Hard skills are the technical and task-based skills that can be taught, practised, and perfected over time. These skills are specific and measurable (Table 1.4).

Table 1.4 Examples of hard skills required by ecologists

Hard skills	*Description*
Species identification	Knowledge and experience in species identification are often required. This implies the ability to identify plants, animals, fungi, or other organisms based on their characteristic attributes.
Data collection, analysis, and interpretation	Ecologists need to collect data – e.g. visual observations (including night vision), acoustic recordings, eDNA and spatial data using manual and remote techniques. They need to handle and analysis vast amounts of data. Therefore, a solid understanding of statistical analysis, processing data, and summarising findings into comprehensible patterns or trends is required. The final stage is to interpret the data which requires knowledge of ecological theory, concepts, and an awareness of other ecological work around the globe.
Geographic Information Systems (GIS) proficiency	Much ecological work involves some spatial aspect. GIS is a powerful mapping technology used to capture, store, manage, and analyse various types of geographical and ecological data. GIS can be used to visualise and interpret data in ways that reveal relationships, patterns, and trends in the form of maps.

There are also new hard skills that are developing as a result of the introduction of new technologies such as eDNA, camera trapping, acoustic monitoring, robotics, and AI (which can be used, for example, to identify birds during night time migration according to their calls in a matter of seconds rather than hours that it would take a human). Each of these technologies is at a different stage of development. These new technologies require an ecologist to combine the hard skills required to use the new technology and interpret the data generated with the soft skills of assessing its usefulness, advantages, and limitations.

1.3.2 INTERDISCIPLINARITY

There is now recognition that many issues cannot be tackled by individuals or groups from one discipline or sub-discipline alone. When added to the need to work 'with' or be led 'by' the local community this leads to a requirement to adopt an interdisciplinary approach to addressing ecological issues.

As an academic movement, interdisciplinarity caught on during the 1970s, since when it has grown. Interdisciplinary research can be thought of as research that approaches an issue from a range of disciplinary perspectives, where the contributions from the various disciplines are acknowledged and integrated to provide a holistic or systemic outcome (Lyall et al. 2011). In this broad sense, interdisciplinarity covers all the ways in which people from different disciplines work together to develop new knowledge and understanding. However, different approaches have been developed which are outlined in Table 1.5.

An example of an interdisciplinary approach can be seen in interventions aimed at conserving a species. In such work it is common to find ecologists working alongside anthropologists, art practitioners, folklorists, geographers, health professionals, hydrologists, planners, psychologists, sociologists, soil scientists together with community leaders and groups, landowners and managers, local and national authorities, and international bodies. The main challenges for the ecologist are developing the soft skills required for teamwork, mutual understanding and trust, and respect for the views of others.

Table 1.5 Definitions of various types of disciplinarity

Type of disciplinarity	*Description*
Cross disciplinary	Viewing one discipline from the perspective of another.
Interdisciplinary	Integrating contributions from several disciplines to a problem or issue by bringing interdependent parts of knowledge together. Therefore, interdisciplinary approaches address the holistic complex of interrelationships.
Intradisciplinary	Work that is undertaken within the same discipline.
Multidisciplinary	Involving several disciplines, each of which provides a different perspective on a problem or issue such that different ideas are combined towards a solution to a common problem.
Transdisciplinary	The unity of intellectual frameworks beyond the disciplinary perspectives. Conceptual frameworks, even though they may have had their origins in one discipline, are adopted across multiple other disciplines as a wider perspective within research and practice.

Source: based on Stember (1991)

1.4 CASE STUDY – ECOLOGICAL EVIDENCE TO POLICY

Modern ecology is an experimental and predictive science which can inform relevant policy (i.e. laws, regulations, directives, and budget allocations put in place by international organisations, governments (national and local), or their representatives). Lawton (2007), based on relevant literature, identified several barriers frequently encountered when incorporating ecological knowledge into policy (Box 1.1). Lawton emphasised that the barriers in his list are not discrete; they overlap or describe different aspects of the same problem.

The process of policy development is an iterative one, with many parties bringing together various interests and options to determine policy. This also means that public policies are continually changing as new issues emerge, attitudes change, the impact of extant policies is monitored and assessed, and new evidence becomes available. This process can be described as a cycle of six steps: agenda

BOX 1.1 BARRIERS TO THE INCORPORATION OF ECOLOGICAL KNOWLEDGE AND UNDERSTANDING INTO POLICY IDENTIFIED BY LAWTON (2007)

1. Scientists do not get the message across clearly enough.
    - This is often called the 'deficit model' (Rayner 2004; Owens 2005). The deficit model takes several forms according to the context.
    - In this context, the deficit model is based on an assumption that the translation of ecological knowledge into policy is linear (identify problem → do science or review literature → formulate policy) and that that policy makers have a deficit of ecological knowledge which the ecologists need to make up by education and the provision of information.
    - The reality is much more complex than this linear model (Sarewitz 2000; Pielke and Rayner 2004). Policy formulation is a messy, iterative, untidy process incorporating economics, cultural values, tensions between institutions, different interpretations of what 'the science' actually tells us, the need to score political points and win political battles, and vested interests.
2. Politicians do not know where to go for the best or most relevant scientific information (Rayner 2002). This is a variant of item 1.
3. The science is ambiguous.
    - Politicians use the uncertainty to avoid difficult decisions (Walton and Gray 1991; Gray 2004; May 2005).
4. Ecologists and policy makers work to different timescales.
    - Ecologists tend to offer advice that is complex and long-term while policy makers want simple short-term solutions (Walton and Gray 1991).
5. 'Institutional failure'
    - Including the wrong decision-making bodies, poor (or no) 'joined-up' government, and contradictory policies in different parts of government (Walton and Gray 1991).

6. The 'paradox of co-operation' (May 2005).
   - Ecologically based solutions frequently require international agreement. The paradox is that unless all nations act together the early adopters may be economically disadvantaged, so no nation wants to be first to act.
7. The scientific advice is contrary to received political wisdom, dogma, or other deeply entrenched beliefs (May 2005).

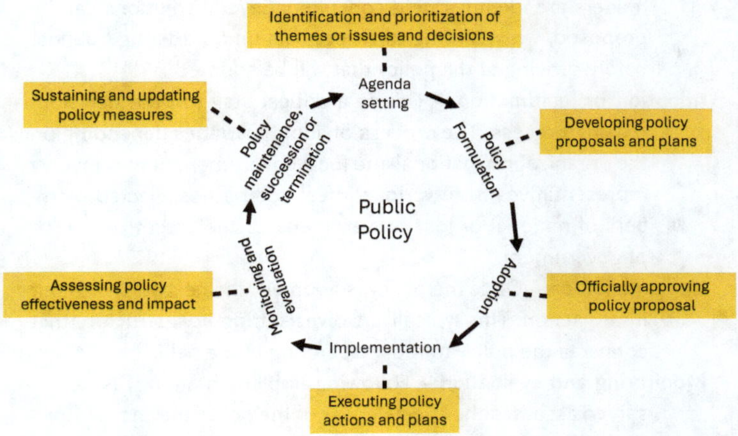

Figure 1.2 The policy cycle.

setting, policy formulation, adoption, implementation, evaluation, and policy maintenance (Figure 1.2). More details of these stages are presented in Box 1.2.

BOX 1.2 DESCRIPTIONS OF THE STAGES IN THE POLICY CYCLE ILLUSTRATED IN FIGURES 1.2 AND 1.3

Agenda setting – In this stage, the need for intervention is identified and lobbied for. There are many international, national, and local organisations as well as individuals who lobby policy makers for policy change. Examining these issues from various perspectives is crucial in deciding what the policy should look like.

Policy formulation – This is the stage that sets up the policy. Objectives are determined, costs are calculated, policy instruments are selected, potential impacts are analysed, and stakeholders are identified and consulted. Several solutions can be proposed, while aiming to meet goals under existing bounds, before arriving at the policy that will be adopted.

Adoption or legitimation – This is a critical stage in the decision-making process. The process of adoption varies depending on the organisation, national and local government. It may involve representatives from various interested parties, elected members of national or local government, or the executive branch of government.

Implementation – Once the policy is adopted, the next step is to put it into action. This typically involves setting up a structure that connects the policy to those operating in the field.

Monitoring and evaluation – Following implementation it is necessary to assesses the effectiveness of the policy they have implemented, and to determine if it has been implemented correctly and has had the desired impact or if there have been unexpected and undesirable consequences. Loopholes may also be identified that will require filling.

Policy maintenance, succession, or termination – This is the point at which decisions are made regarding keeping the policy as it is, changing it in the light of new evidence, unexpected consequences, changing attitudes, etc., or ending that the policy.

Relevant ecological knowledge and understanding can be fed into the policy cycle at almost every stage (Figure 1.3).

Policy combines knowledge and values, and it is the values that tend to influence how information is interpreted. This emphasises the importance of recognising and addressing commonly held values among individuals and groups. Using multi-stakeholder dialogues can help ecologists and policy makers communicate more effectively with each other and with those outside of these professions. To do so requires ecologists to draw on their soft skills, particularly their communications skills (Figure 1.4). Of course, they need to base their communication on solid evidence and on their

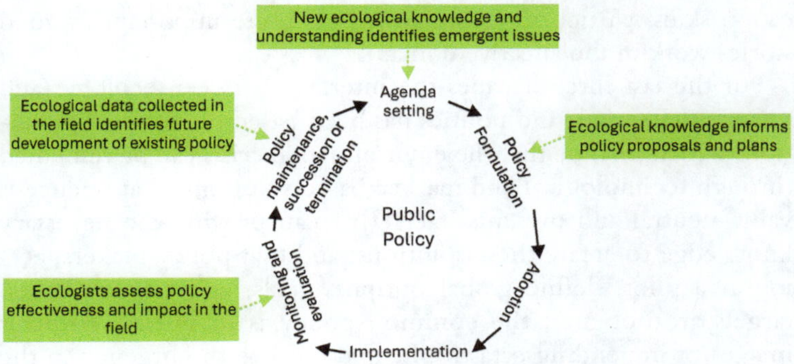

Figure 1.3 Possible points of input of ecological knowledge and understanding into the policy cycle.

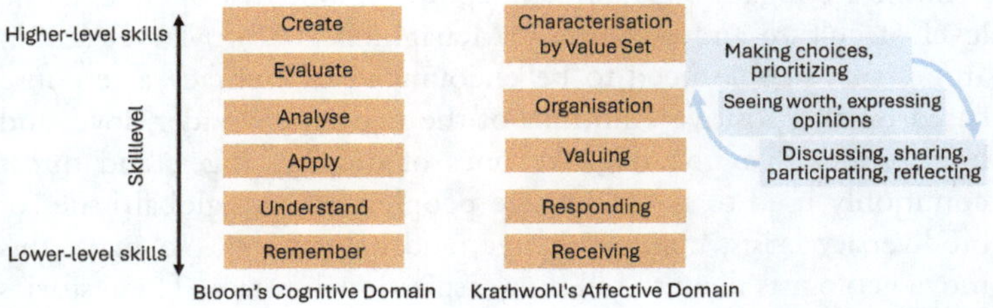

Figure 1.4 The cognitive and affective domains.

knowledge of ecology. Science is primarily based in the cognitive domain and, hence, is how scientists tend to talk especially to each other and to others. In this domain, evidence and facts are important. The affective domain describes how knowledge is integrated into a person's frame of reference and in a social context. The affective domain is the domain in which learning is contextualised or situated. This is the domain in which stories become important. Stories shape the world view of people, developing their personal set of values and behaviours. A fact may be presented cognitively but the response will be mediated by a person's set of values. The way that information is presented is important. It needs to be done in a way that people can receive it easily, and in a way that they can respond to based and building on their values. Good communication skills, particularly story telling skills are important as good stories work in the affective domain.

For the last three decades the interface between ecology (and science in general) and politics has been based on a set of convictions which include that the environmental crisis can be remedied through technological and market-based solutions; that science is value-neutral and presents 'facts' that can provide the necessary knowledge to create these solutions; and that policy makers, acting as a pilot, define global options, goals, and targets. When targets are not met, the common policy is to present stronger images of impending catastrophe and escalate the urgency in the language that is used (Mcnaghten 2025). The outcome is, generally, to further exacerbate a sense of despair, despondency, and alienation and to diminish citizens' feelings of political agency (Mcnaghten 2025).

Some are suggesting a different approach, one that operates at the level of culture and belonging (Mcnaghten 2025). Marvel (2025) argues that people need to be encouraged to embrace a response based on the positive emotions of the need for wonder, love, and hope rather than the negative ones of despair, anger, and threat commonly used to try to engage people with the global issues of biodiversity crisis, climate change, and habitat loss. To make this move ecologists need to tell more aspirational stories. These stories should give hope and renewal to raise awareness of potential futures (Mcnaghten 2025).

1.5 SELF-STUDY SUGGESTIONS

1.5.1 SELF-STUDY QUESTIONS

1. Define ecology
2. Why are there so many branches of ecology?
3. What is the difference between a qualitative and a predictive science?
4. What is the difference between ecology and natural history?
5. What makes ecology an empirical science?
6. What is a hypothesis?
7. How do the inductive and deductive approaches to testing hypotheses differ? What are the strengths and weaknesses of each approach?
8. What distinguishes a dependent variable from an independent variable?
9. What is the importance of replication and controls in experiments?
10. What is a model in an ecological sense?
11. What is the role of ecology in multi- and interdisciplinary programmes?
12. How do science and policy interact?
13. What do policy makers need to know from scientists and why is it often difficult for scientists to provide that information when asked for it?
14. What skills does an ecologist have?
15. What are the current major global environmental issues and how does a knowledge and understanding of ecology help to address these?

1.5.2 A DEEPER DIVE

A. In 2018 Courchamp and Bradshaw (2018) published a list of 100 articles every ecologist should read. Download a copy of that paper and then read some or all of the 100 papers they suggest.

B. Explore websites for the WWF, IUCN, United Nations Convention on Biological Diversity, and national and regional government and related organisations that produce policy for your country.

C. Visit the website of the Chartered Institute of Ecology and Environmental Management (https://cieem.net) to look at the skills required by an ecologist. What skills do you need to gain and how will you develop those skills?
D. What data-analysis packages do you have available to support your studies?
E. Visit www.scienceofpeople.com/ which is a site providing guidance on improving communication skills.
F. If you wish to develop some practical field skills, consider reading Karban et al. (2023) and Wheater et al. (2020).

1.6 REFERENCES

Beltran, R.S., Marnocha, E., Race, A., Croll, D.A., Dayton, G.H. and Zavaleta, E.S. (2020). 'Field courses narrow demographic achievement gaps in ecology and evolutionary biology'. *Ecology and Evolution* 10(12): 5184–5196. https://doi.org/10.1002/ece3.6300.

Commoner, B. (1971). *The Closing Circle: Nature, Man, and Technology*. New York: Knopf.

Courchamp, F. and Bradshaw, C.J.A. (2018). '100 articles every ecologist should read'. *Nature Ecology & Evolution* 2(2): 395–401. https://doi.org/10.1038/s41559-017-0370-9.

Demery, A-J.C. and Pipkin, M.A. (2021). 'Safe fieldwork strategies for at-risk individuals, their supervisors and institutions'. *Nature, Ecology & Evolution* 5(1): 5–9. https://doi.org/10.1038/s41559-020-01328-5.

Gray, A.J. (2004). 'Ecology and government policies: the GM crop debate'. *Journal of Applied Ecology* 41: 1–10. https://doi.org/10.1111/j.1365-2664.2004.00873.x.

Haeckel, E.H.P.A. (1866). *Generelle Morphologie der Organismen. Allgemeine Grundzüge der organischen Formen-Wissenschaft, mechanische Begründet durch die von Charles Darwin reformirte Descendenz-Theorie*. 2 vols. Georg Reimer: Berlin, Germany.

Karban, R., Huntzinger, M. and Pearse, I.S. (2023). *How to Do Ecology: A Concise Handbook* (3rd edition). Princeton, NJ: Princeton University Press.

Lawton, J.H. (2007). 'Ecology, politics and policy'. *Journal of Applied Ecology* 44: 465–474. https://doi.org/10.1111/j.1365-2664.2007.01315.x.

Lyall, C., Bruce, A., Tait, J. and Meagher, L. (2011). *Interdisciplinary Research Journeys: Practical Strategies for Capturing Creativity*. London: Bloomsbury Academic. http://doi.org/10.5040/9781849661782.

Marvel, K. (2025). *Human Nature: Nine Ways to Feel about Our Changing Planet*. New York: Ecco Press.

May, R.M. (2005). 'Threats to tomorrow's world: Anniversary Address 2005'. *Notes and Records of the Royal Society*, 60: 109–130. https://doi.org/10.1098/rsnr.2005.0134.

Mcnaghten, P. (2025). 'Environmental politics is doomed to fail – unless we tell better stories'. *Nature* 644, 32–34. https://doi.org/10.1038/d41586-025-02459-0.

Owens, S. (2005). 'Making a difference? Some perspectives on environmental research and policy'. *Transactions of the Institute of British Geographers* 30: 287–292 [Commentary]. www.jstor.org/stable/3804406.

Pielke, R.A. Jr and Rayner, S. (eds). (2004). 'Science, policy and politics: learning from the controversy over *The Skeptical Environmentalist*'. *Environmental Science & Policy* 5: 355–434.

Rayner, S. (2002). 'We know enough'. *Guardian*, 2 September.

Rayner, S. (2004). 'The novelty trap: why does institutional learning about new technologies seem so difficult?'. *Industry and Higher Education*, December, 349–355.

Sarewitz, D. (2000). 'Science and environmental policy: an excess of objectivity'. In R. Frodeman (ed.), *Earth Matters: the Earth Sciences, Philosophy, and the Claims of Community*, 79–98. Upper Saddle River, NJ: Prentice Hall.

Stember, M. (1991). 'Advancing the social sciences through the interdisciplinary enterprise'. *The Social Science Journal* 28(1): 1–14. https://doi.org/10.1016/0362-3319(91)90040-B.

Walton, D.W.H. and Gray, A.J. (1991). 'Ecology and government policies'. *Trends in Ecology and Evolution* 6: 144–145. https://doi.org/10.1111/j.1365-2664.2004.00873.x.

Wheater, C.P., Bell, J.R. and Cook, P. (2020). *Practical Field Ecology: A Project Guide*. Oxford: John Wiley & Sons.

2

THE ABIOTIC (NON-LIVING) ENVIRONMENT

2.1 INTRODUCTION

Ecologically, the environment is composed of living and non-living elements. The living elements include bacteria, fungi, plants, and animals. The word 'biotic' (from the Greek 'biotic' meaning 'living'). These elements will be the focus of Chapters 3–7 in this book. The non-living (i.e. 'abiotic' – derived from the Greek 'a' meaning 'non' and 'biotic' meaning 'living') elements of the environment can be divided into chemical and physical factors (Figure 2.1).

Abiotic factors are important components of the structure and function of the environment and govern the fundamental processes that allow life to exist. The chemical and physical properties of soil, (particularly pH, mineral content, and texture) are highly influential in determining which bacterial, fungal, invertebrate, and vertebrate species live (or burrow) in the soil and which plants can grow and thrive. Temperature directly regulates the metabolic function of individual organisms. Light drives the flow of energy through an ecosystem and regulates the behaviour of many aquatic and terrestrial plants and animals. Precipitation and temperature are major factors in determining the global distribution of life determining the natural vegetation of a given area (e.g. tropical rainforest, cold tundra, or dry desert). Variations in abiotic factors can be the result of natural processes or human activities.

The focus in this chapter is to develop a foundational knowledge of abiotic factors required by ecologists. For example, ecologists may be required to assess the ecological quality of a woodland. To do so they need to know what species might be expected in that

DOI: 10.4324/9781003604464-2

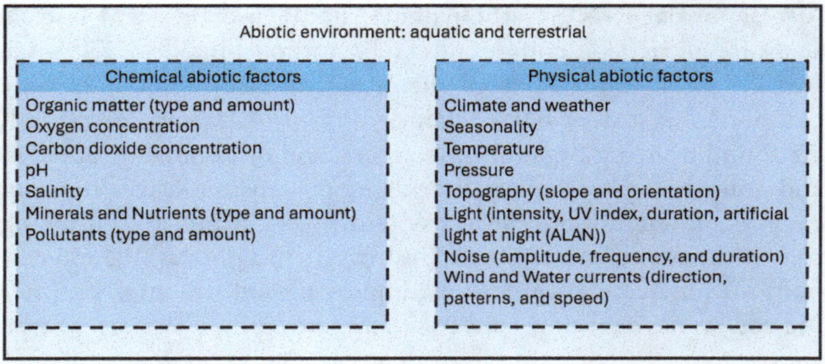

Figure 2.1 Examples of abiotic factors.

location. That requires an understanding of the local abiotic conditions for example, the climate (including temperature and rainfall patterns through the year), and the soil type and condition. It is often the case that on entering a new area, the first thing an ecologist does is to look at the ground and examine the soil (i.e. they are looking at some of the abiotic factors specific to that location).

Understanding abiotic factors is essential for ecological research, conservation efforts, and managing natural resources effectively. By studying these factors, scientists can better understand how ecosystems function and how to protect them from environmental changes and human impacts. The second and third sections in the chapter contain descriptions of important abiotic factors and their relevance to ecologists when addressing issues such as environmental change.

The case study highlights the ecological implications of a changing climate. The case study ends with an overview of strategies that can be used to address the ecological impacts of climate change, many of which will be discussed in greater detail in later chapters in this volume.

2.2 CHEMICAL FACTORS

2.2.1 ORGANIC MATTER

Organic compounds contain carbon atoms (C) bonded with hydrogen (H) and often with other elements, most commonly oxygen

(O) or nitrogen (N). Carbohydrates, lipids, proteins, and nucleic acids are all organic compounds (n.b., carbon dioxide (CO_2 – see Box 2.1 for an explanation of chemical formulae) is not an organic compound as it does not contain hydrogen). Organic compounds are found in living organisms, in faeces, and dead remains of plants and animals (i.e. organic matter). Organic matter drives nutrient cycling and energy flow in ecosystems (see Chapter 6); enriching soils with essential nutrients (e.g. nitrogen, phosphorus (P), and carbon) thus increasing soil fertility, improving soil structure, enhancing soil water retention, and mitigating erosion. These are critical factors for both agricultural productivity and ecosystem resilience (see Chapter 6).

BOX 2.1 UNDERSTANDING CHEMICAL FORMULAE

Elements

An element is a chemical substance that cannot be broken down into a simpler substance. Each chemical element has been assigned a code of one or two letters. Examples of elements mentioned in this book include hydrogen (H), carbon (C), chlorine (Cl), nitrogen (N), oxygen (O), phosphorous (P), potassium (K), selenium (Se), sodium (Na), and sulphur (S).

Molecules

A molecule comprises two or more atoms of the same or different element. The number of each atom of each element in a molecule is shown by a subscript number after the symbol for the element. For example, oxygen in the atmosphere occurs as a molecule comprising two oxygen atoms; hence the symbol for this oxygen molecule is O_2. Other molecules mentioned in this book include:

- Water – H_2O – two hydrogen atoms and one oxygen atom.
- Calcium carbonate – $CaCO_3$, – one atom of calcium, one atom of carbon, and three atoms of oxygen.

- Calcium bicarbonate – $Ca(HCO_3)_2$ – one atom of calcium, two atoms of hydrogen, two atoms of carbon, and six atoms of oxygen.
- Carbon dioxide – CO_2 – one carbon atom and two oxygen atoms.
- Carbonic acid – H_2CO_3 – two atoms of hydrogen, one atom of carbon, and three atoms of oxygen.
- Glucose – $C_6H_{12}O_6$ – six carbon atoms, 12 hydrogen atoms, and six oxygen atoms.
- Hydrogen sulphide – H_2S – two hydrogen atoms and one sulphur atom.
- Nitrogen oxides – NO_x – one nitrogen atom and x oxygen atoms where x is typically one or two. This is a group of nitrogen oxides (in the press and online often referred to as NOX), primarily nitric oxide (NO) and nitrogen dioxide (NO_2), which are significant air pollutants.
- Magnesium carbonate – $MgCO_3$ – one atom of magnesium, one atom of carbon, and three atoms of oxygen.
- Magnesium bicarbonate – $Mg(HCO_3)_2$ – one atom of magnesium, two atoms of hydrogen, two atoms of carbon, and six atoms of oxygen.
- Methane – CH_4 – one carbon atom and four hydrogen atoms.
- Sodium carbonate – Na_2CO_3 – one atom of sodium, one atom of carbon, and three atoms of oxygen.
- Sodium bicarbonate – $NaHCO_3$ – one atom of sodium, two atoms of hydrogen, two atoms of carbon, and six atoms of oxygen.
- Sodium chloride (common salt) – NaCl – one sodium atom and one chlorine atom.
- Sulphur dioxide – SO_2 – one carbon atom and two oxygen atoms.
- Potassium carbonate – K_2CO_3 – two atoms of potassium, one atom of carbon, and three atoms of oxygen.
- Potassium bicarbonate – $KHCO_3$ – one atom of potassium, one atom of hydrogen, one atom of carbon, and three atoms of oxygen.

In the press and online carbon dioxide, sulphur dioxide and nitrogen oxides are often referred to as CO2, SO2, and NOX (note the absence of subscripts). The ecologists should recognise that these are abbreviations for the chemicals and not their chemical formula.

Where a number occurs in front of the symbol for a molecule, this indicates the number of molecules. Where there is no number this indicates that one molecule is involved in the reaction. The equation for photosynthesis, for example is:

$$6CO_2 + 6H_2O \rightarrow C_6H_2O_6 + 6O_2$$

In words this equation is read as:

Six molecules of carbon dioxide + six molecules of water → one molecule of glucose + six molecules of oxygen

Ions

An atom comprises a nucleus in which there are neutrons which carry no electrical change and proton which carry a positive charge. Outside the nucleus there are electrons which carry a negative change. The charge carried by one proton and one electron are identical, but opposite. The simplest element, hydrogen has one proton and one electron and so an atom of hydrogen, in common with all atoms, is overall neutral in terms of the charge it carries. An atom of oxygen has eight proton and eight electrons and is overall neutral in terms of charge. In a molecule the overall charge is also zero. In some circumstances an atom, alone or as part of a molecule, may lose or gain one or more electrons to form an ion. If electrons are lost, the ion will have more protons than electrons and will be positive charged (termed a cation), and when electrons are gained, the ion will have more electrons than protons and the ion will be negatively charge (termed an anion). The charge carried by an ion is shown as either a + or – as a superscript after the symbol for the atom or molecule. Examples of ions mentioned in this book include:

- Ammonium ions – NH_4^+ – one nitrogen atom and four hydrogen atoms with an overall charge of +1.
- Calcium ions – Ca^{2+} – one atom of calcium that has lost two electrons and, hence, the resultant ion has an overall charge of +2.
- Hydrogen ions – H^+ – one atom of hydrogen that has lost one electron and, hence, the resultant ion has an overall charge of +1.
- Hydroxide ions – OH^- – one atom of oxygen and one atom of hydrogen with an overall charge of –1.
- Magnesium ions – Mg^{2+} – one atom of magnesium that has lost two electrons and, hence, the resultant ion has an overall charge of +2.
- Nitrate – NO^{3-} – an ion comprising one nitrogen atom and three oxygen atoms carrying an overall charge of –1.
- Nitrite – NO^{2-} – an ion comprising one nitrogen atom and two oxygen atoms carrying an overall charge of –1.
- Potassium ions – (K^+) – one atom of potassium that has lost one electron and, hence, the resultant ion has an overall charge of +1.
- Sodium ions – (Na^+) – one atom of sodium that has lost one electron and, hence, the resultant ion has an overall charge of +1.

2.2.2 OXYGEN AND CARBON DIOXIDE CONCENTRATION

The concentrations of oxygen and carbon dioxide in the atmosphere at sea level are, respectively, around 20 per cent and 0.04 per cent. Due to the mixing effect of air currents, these concentrations are relatively consistent across the globe. However, there are important variations. Volcanic activity may cause localised build-up of carbon dioxide, which can lead to the death of animals that venture into these areas. More significant on a global scale is the increase in atmospheric carbon dioxide associated with the burning of fossil fuels. Prior to the Industrial Revolution of the late eighteen and early nineteenth centuries atmospheric carbon dioxide levels were around 280 ppm (ppm = parts per million) (0.028 per cent) and today the level is 415 ppm (0.0415 per cent).

In comparison to the atmosphere, the aquatic environment is much more variable in terms of the concentration of oxygen and carbon dioxide. The amount of any gas that water can hold depends on temperature, salinity, and pressure. The solubility of a gas in water decreases as temperature increases and increases with both salinity and pressure. Thus, cold water holds more gas than warm water; saline water holds more oxygen than freshwater, and water at greater depth can hold more oxygen than that nearer the surface. Generally, dissolved oxygen in water is around 0.0008–0.0014 per cent (i.e. 8–14 mg l^{-1} (mg l^{-1} = milligrams per litre). However, oxygen levels tend to be highest at the water surface, where the water absorbs oxygen from the atmosphere, and decreases with increasing depth because of the aerobic decomposition of material (which uses oxygen) on and in the bottom sediment. In some very clear ponds and lakes there is little variation in oxygen levels with depth as sunlight can penetrate further into the water than in turbid water (i.e. water that is cloudy) allowing green plants to photosynthesise releasing oxygen into water at the bottom of the waterbody.

Dissolved oxygen levels less than 5 mg l^{-1} are considered stressful for fish, and few species of fish can live in water with less than 3 mg l^{-1} oxygen. Dissolved oxygen levels below 1 mg l^{-1} are considered hypoxic (i.e. inadequate to maintain life) (US EPA 2023). Oxygen is lost from water as temperature increases, as it is used by plants and animals in respiration, and through aerobic decomposition of organic matter. In summer, as the water warms, oxygen is lost from the water, a situation that is exacerbated if air pressure is low. As a result of warm temperatures, either alone or in combination with low atmospheric pressure, oxygen levels can fall below that at which fish can survive.

In temperate climates water circulates throughout the water column during spring and autumn overturns (for an explanation of this season shift, see section 2.3.3 and Figure 2.11). Overturns replenish oxygen at lower levels in the water column. In winter, oxygen levels in unfrozen water bodies tends to remain constant as the water is colder, and so can hold more oxygen than warm water, and there is reduced decomposition. However, under ice, oxygen levels can fall and can lead to fish death.

When water is agitated by passing over rocks that break the water surface (e.g. where it cascades, or where wind stirs up the surface of water) the surface area increases and more oxygen can diffuse into the water. In shallow, rapidly flowing water and in wind-driven sprays (i.e. where there is a large surface area over which oxygen can be absorbed) oxygen levels can be saturated and even supersaturated. Oxygen also enters water from gas released by photosynthesising submerged water plants.

The main source of carbon dioxide in water is the aerobic and anaerobic decomposition of organic matter which is composed of the remains of aquatic plants and animals and of terrestrial plants and animals that are carried into water (e.g. by flood water). Carbon dioxide is present in water as free CO_2 (i.e. as a gas) or it can be dissolved. When carbon dioxide dissolves in water it reacts with water to produce carbonic acid (H_2CO_3).

2.2.3 pH

The pH of an environment is described as acidic (pH < 7 – i.e. a pH less than 7), basic/alkaline (pH > 7 – i.e. a pH greater than 7), or neutral (pH is exactly 7). pH impacts a wide variety of chemical and biological processes including the solubility and toxicity of pollutants, the availability of nutrients, and the physiological functions of aquatic organisms. Most living organisms function best in a pH environment that is broadly neutral (i.e. the pH is 7 or very close to 7).

Rainwater is naturally slightly acidic due to the absorption of carbon dioxide from the atmosphere and the formation of carbonic acid. Pollutants such as sulphur dioxide (SO_2) and nitrogen oxides (NO_x), can significantly lower the pH of rainwater (i.e. making it more acidic – so called 'acid rain'). In addition to these atmospheric factors, several other factors can influence pH in aquatic environments, for example, organic matter, photosynthesis, pollutants and the bedrock. Decaying organic matter from vegetation can release acids into the water, lowering the pH. During photosynthesis, aquatic plants and algae consume carbon dioxide, which can raise the pH of the water (i.e. more alkaline). Industrial and agricultural pollutants can also introduce acids or bases into aquatic environments. For example, a limestone bedrock (which is alkaline) can buffer against acidification (i.e. prevent changes in the pH meaning that the pH of the water remains near neutral – around pH 7–8.5). Granite bedrock provides little buffering capacity and the water flowing over it can become acidic.

In the terrestrial environment, soils can be divided into those that are acidic (with a pH below 7), those that are neutral (with a pH around 7), and those which are alkaline (with a pH above 7). Soil pH reflects the balance between hydrogen ions (H^+) and hydroxide ions (OH^-) in the soil solution, The levels of these ions is controlled by the weathering of the parent rock, leaching, and biological activity. In humid regions, heavy rainfall leaches basic cations (i.e. positively charged ions) such as calcium (Ca^{2+}), magnesium (Mg^{2+}), potassium (K^+), and sodium (Na^+) from the soil, while the decomposition of organic matter releasing organic acids. Collectively, leaching and release of organic acids, lowers pH and produces acidic soils. Neutral soils develop where inputs of basic cations, from mineral weathering and atmospheric deposition, balance acid production to maintain a pH around 6.5–7.5. In arid and semi-arid environments there is limited leaching, which allows calcium, magnesium, potassium and sometimes sodium carbonates (i.e, $CaCO_3$, $MgCO_3$ K_2CO_3, and Na_2CO_3) and bicarbonates (i.e. $Ca(HCO_3)_2$, $Mg(HCO_3)_2$, $KHCO_3$, and $NaHCO_3$) to accumulate and elevate pH above 7, (i.e. alkaline soils).

2.2.4 SALINITY

Salinity is a measure of the amount of salt – primarily sodium chloride (NaCl) – dissolved in the water (Table 2.1).

Table 2.1 Salt content (percentage) of aquatic environments. The salt concentration in water is frequently given in parts per thousand (‰) – to convert from per cent (%) to parts per thousand (‰) multiple the percentage by ten (e.g. 3.5 per cent is the same as 35 parts per thousand)

Aquatic environment	*Salt content (%)*
Freshwater	< 0.05 (typically)
Brackish water	0.05–3
Salt (saline) water	3–5 (with a mean of around 3.5 per cent but with variation across the globe)
Brine (forms naturally due to evaporation of saline water)	> 5 – up to 26–28 maximum

Osmosis describes the movement of water across a semi-permeable membrane from a solution that is hypotonic (more water, less solute) to a hypertonic solution (less water, more solute). One of the implications of this is that, under the force of osmosis, water tends to flow into a plant or animal living in freshwater (i.e. the plant or animal contains less water and more solute (i.e. it is hypertonic) compared to the freshwater) while in salt water the water tends to flow out of the organism (i.e. the plant or animal contains more water and less solute (i.e. it is hypotonic) compared to the saltwater. This has implications for the physiology of plants and animals.

2.2.5 MINERALS AND NUTRIENTS IN WATERBODIES

Minerals are the chemical elements in the Earth's crust. They are released from bedrock through natural processes, such as physical, chemical, and biological weathering, and their concentrations in soils are affected by anthropogenic activities such as mining, energy production, industrial and agricultural activities, and waste disposal (Brown and Calas 2011). Some minerals are classed as nutrients (i.e. they are essential for the growth of plants and animals), while others are contaminants (any substance present in an environment where it does not normally occur or is present in higher concentrations than normal) and pollutants (a specific type of contaminant that has been introduced into an environment and is present beyond permissible limits and causes measurable harm) affecting plant and animal health. Depending on the amount taken up or ingested, some minerals can be both essential and harmful to plant and animal life. Selenium (Se), for example, is an essential trace mineral for animals but harmful (toxic) in high quantities. Minerals contained in bedrock and those released during industrial and agricultural activities dissolve in water while those in soil may be blown or carried into a waterbody (a stream, river, canal, pond, or lake) due to erosion.

Waterbodies are described according to the level of nutrients they contain (Table 2.2).

2.2.6 SOILS

Soils contain minerals and nutrients. As with the aquatic environment, these are largely derived from the bedrock which, being

Table 2.2 Classification of waterbodies according to nutrient content

Classification	*Nutrient content*	*Description*
Oligotrophic	Nutrient poor (typically low levels of nitrogen and phosphorus) and low productivity	Usually found in the cold regions. Mixing of nutrients is rare and slow due to the low water temperature. Tend to have very clear water and few plants and support aquatic species requiring well-oxygenated, cold waters.
Mesotrophic	Intermediate levels of nutrient and productivity	Usually clear water with submerged aquatic plants.
Eutrophic	High levels of nitrogen and phosphorous and are frequently highly productive	High nutrient levels combined with high levels of oxygen initially results in the accelerated multiplication and growth of lake flora and fauna. Plants or algal blooms can block the sun from penetrating the waterbody. This leads to reduced oxygen levels due to the increased levels of respiration. Fauna suffers and may not be able to survive in the dark, low oxygen water. Eutrophication might occur naturally or be due to industrial, agricultural, or other anthropogenic activities.
Hypereutrophic	Very high levels of nutrient and productivity	Excessive plant and algal growth, due to a high supply of growth nutrients, result in reduced water transparency (usually less than 1 metre). This might create dead zones beneath the water surface in which fauna suffocates due to low oxygen levels.

varied, results in different soil types, each with distinctive characteristics (Table 2.3 and Figure 2.2).

Pollutants in soil (and in air and water) can be both naturally forming (i.e. minerals or extracted compounds like oil) or anthropogenic in origin (i.e. manufactured materials or byproducts).

Table 2.3 Soil types

Soil type	*Characteristics*
Sandy soil	• Feels gritty • Develop over limestone or chalk bedrock rich in calcium carbonate ($CaCO_3$) • pH > 7.5 (alkaline) • Drainage – tends to be free draining, so retains little water • Nutrients other than calcium are limited
Clay soil	• Sticky when wet and hard when dry, retains water • Roots find it difficult to penetrate clay soils due to the tight packing of the small soil particles • pH 6.0–8.0 depending on the organic matter and mineral content • Drainage– tends to be poor so retains water • Nutrient retention: high – generally rich in essential macronutrients because they are held strongly on clay surfaces
Silty soil	• Feel slippery when wet and smooth and floury when dry • Holds moisture better than a sandy soil • Easily compacted • pH ranges from 6 to 7.5 (slightly acidic to neutral) • Drainage – well drained and water retention is between than in sandy soils • Nutrient retention – holds nutrients more effectively than sand but less so than clay • Compaction – easily compacted
Loam	• Feels soft and crumbly • Drains well • Mixture of sand (40 per cent), silt (40 per cent), and clay (20 per cent) • pH 6.5–7.5 (neutral) • Drainage – good but also retains moisture • Nutrient retention – balanced – typically, moderate to high cation exchange capacity (CEC) due to the clay and organic matter content

Source: adapted from Anon (2025)

Pollutants may cause long- or short-term damage due to their lethal and sub-lethal effects. Sub-lethal effects include changed growth rate and reproductive capabilities. Some pollutants persist in the environment and are called 'forever chemical'. One such example is PFAS (per- and polyfluorinated alkyl substances) used in paper and cardboard food packaging (e.g. takeaway containers, popcorn bags,

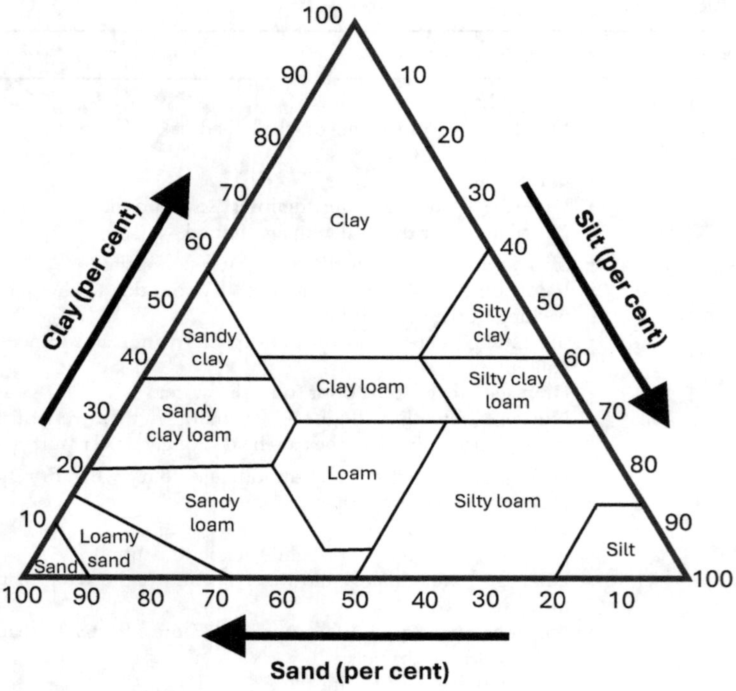

Figure 2.2 Soil triangle.

pizza boxes, ready-made cake boxes etc.), non-stick cookware, textiles (e.g. waterproof outdoor clothing and equipment, and carpets, mattresses), cosmetics (e.g. hair conditioner, foundation cream, and sunscreen), and electronics (e.g. smartphones) which hardly degrade in the natural environment.

An important issue with some soils is compaction in which the soil's network of pores collapses. The network of pores normally holds air and water but when compacted, which may be the result of livestock trampling, foot traffic, and heavy traffic, the pore structure collapses. The effect of this is that water infiltration slows, surface runoff and erosion accelerate, oxygen diffusion to roots and soil microbes is curtailed, and plant roots are less able to penetrate the soil (which reduces access nutrients). Where compaction leads to anaerobic pockets (i.e. no oxygen is present) there is a shift in the composition of microbial communities. Over time, these physical, chemical, and biological disruptions cascade through food webs (see Chapter 6) favouring disturbance-tolerant species (see Chapter 4, section 4.5) and eroding ecosystem resilience (see Chapter 6).

2.3 PHYSICAL FACTORS

2.3.1 CLIMATE AND WEATHER

The climate is controlled by many factors (Figure 2.3). These factors interact with each other to produce the different climates around the world.

Climate and weather variables include temperature, precipitation, humidity, hours of sunshine, cloud cover, atmospheric pressure, fogginess, and wind velocity. Weather is the term used for short-term changes in these variables, usually measured in hours or days. Climate is the average, extremes, frequency, and intensity of all the weather that takes place over a long period of time, usually years or decades. The weather and climate of a particular location are affected by latitude, longitude, terrain, altitude, land use, and nearby waterbodies and their currents (Gough and Leung 2022).

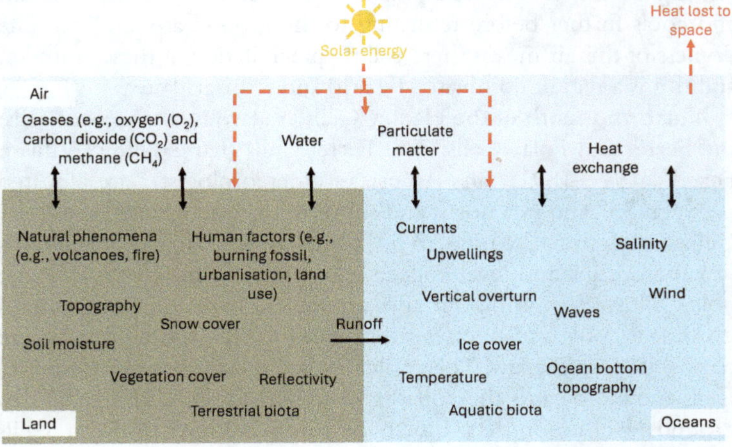

Figure 2.3 Simplified climate model indicating some of the factors that interact and control the climate.

2.3.1.1 *Air movements*

As the Earth spins it causes the air mass above the Earth's surface to spin in the same direction. Winds starting at the equator move north at a height of about 12–15 km (they are termed south winds as they are blowing from the south). The Coriolis effect (named, in 1835, after Gaspar-Gustave de Coriolis (1792–1843), a French mathematician and engineer) causes winds in the Northern Hemisphere to be deflected clockwise and anticlockwise in the Southern Hemisphere. Frictional drag as the winds move over the Earth's surface causes them to slow and, hence, air pile up at about 25° north. The same happens south of the equator except that air is deflected to the west rather than the east. There are therefore two atmospheric cells, one north and the other south of the equator. These cells are called the Hadley Cells (also known as Hadley circulation – named after George Hadley (1685–1768, an English lawyer and amateur meteorologist) who put forward the idea in 1735 even though it was Edmund Halley (1656–1742, an English astronomer, mathematician, and physicist) who, first proposed their existence 50 years earlier in 1685. In the Hadley Cells, warm air rises at the equator (Figure 2.6). As it gains altitude, the air cools and the water vapour it is carrying turns to precipitation. It is this precipitation that gives rise to tropical rainforests. Having lost much of its water, the now drier air moves north or south and cools further before returning to the ground around 25°. The dryness of this air means there is little precipitation at these latitudes, and this is a region dominated by arid and semi-arid areas.

North and south of the Hadley Cells are two further sets of cells: the Ferrel and Polar Cells. The Ferrel Cells (named after William Ferrel (1817–1891, an American meteorologist) are located between 35° and 60° north and south of the equator and the Polar Cells stretch from the poles to 60° north or south In the Ferrel Cells warm air at ground level collects moisture as it travels towards the poles. This warm, moist air rises around 60° north or south and as it does, it cools and releases precipitation. This is the zone where temperate forests grow. The cooler air, at high altitude, in the Ferrel Cells moves towards the equator as cooler dry air and descends to ground level around 25° north or south Within the Polar Cells cold air at ground level flows away from the poles. When this air reaches 60° it has warmed sufficiently to rise in the atmosphere where it moves back towards the poles. As the air in the Polar Cells is cold, it carries little moisture and Polar regions are relatively dry (Figure 2.6).

The result of these atmospheric cells are areas around the equator dominated by rain forests. To the north and south of these rain forests are areas of dessert. Further north and south there are areas of temperature woodland and grasslands, and towards the poles there are boreal forest, tundra, and ice caps in the north and ice caps in the south (Figure 2.4).

Global weather patterns can be influenced by jet streams. These are fast flowing, narrow air currents in the Earth's atmosphere. The Northern and Southern Hemispheres each have a polar jet around 9.1 km above sea level and typically travelling at around 180 km h^{-1} (although often considerably faster) around their respective polar vortex (more formally a circumpolar vortex – a large region of cold, rotating air which encircles the Earth's polar regions) (McCabe 2022). The northern polar jet flows west to east over the middle to northern latitudes of North America, Europe, Asia, and their intervening oceans, while the Southern Hemisphere polar jet stream mostly circles Antarctica also from west to east. There are also somewhat higher and weaker subtropical jet streams located north and south of the equator (McCabe 2022). As with the Polar jet streams the subtropical jets streams also flow west to east.

2.3.1.2 Water currents

As winds move across oceans, they cause water currents. These currents transport water, minerals and sensible heat (i.e. the exchange of heat which is seen in a change in temperature) around the globe. There are two main oceanic systems: the Atlantic and Pacific. Both oceans range from pole to pole and are bordered on either side by extensive land masses that deflect ocean currents. The currents in each ocean are dominated by two large circular water motions known as gyres. In each ocean there is one gyre to the north of the equator and one to the south, both centred around the tropical and sub-tropical regions. Due to the Coriolis effect, the currents in the Northern Hemisphere circulate clockwise and those in the Southern Hemisphere anticlockwise. This circulation pattern is also influenced by the prevailing wind directions: the trades or tropical easterlies near the equator and the westerlies on the pole side. There is an easterly flowing equatorial current that separates the gyres (Figure 2.5).

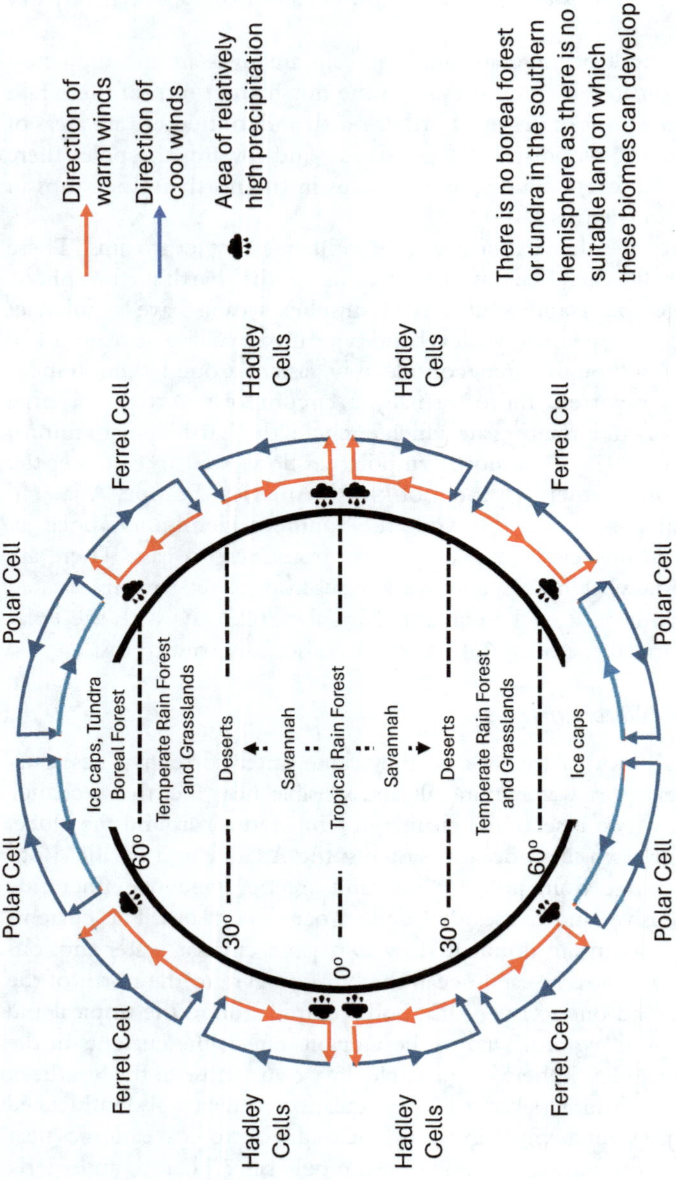

Figure 2.4 Atmospheric cells, wind directions, and resultant vegetation patterns.

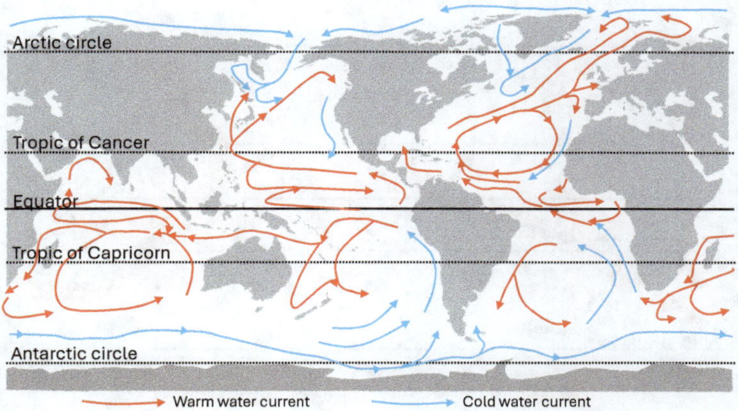

Figure 2.5 Warm and cold ocean surface currents.

Source: base map Public Domain, https://commons.wikimedia.org/w/index.php?curid=868126 – the Arctic Circle, Tropic of Cancer, Equator, Tropic of Capricorn, and Antarctic Circle, Antarctic land mass and currents have been added to the base map

Cold water being more dense than warm water sinks and travels at greater depths than warm water. The overall effect of this is a global circulation of water (Figure 2.6).

The El Niño–Southern Oscillation (ENSO – a climate phenomenon marked by the unusual warming of surface waters in the central and eastern tropical Pacific Ocean), its counterpart is La Niña (marked by unusually cold sea surface temperatures in the central and eastern Pacific Ocean, driven by stronger-than-normal trade winds) affect the location of the jet streams, which in turn influences the weather over the tropical Pacific Ocean, much of the tropics and subtropics, and higher-latitude regions.

As currents encounter the continents, they rise. This is called an upwelling (Figure 2.7). The rising cold water brings with it a supply of nutrients from the seabed which supports a rich food web of phytoplankton, zooplankton, macroinvertebrates, fish, birds, and mammals. Being cold, these upwelling currents transfer little moisture into the atmosphere. Hence, the onshore winds are dry, and the result is a series of deserts along the coast of Mexico and South America and chaparral (a shrubland community of plants) in coastal regions of California and Chile.

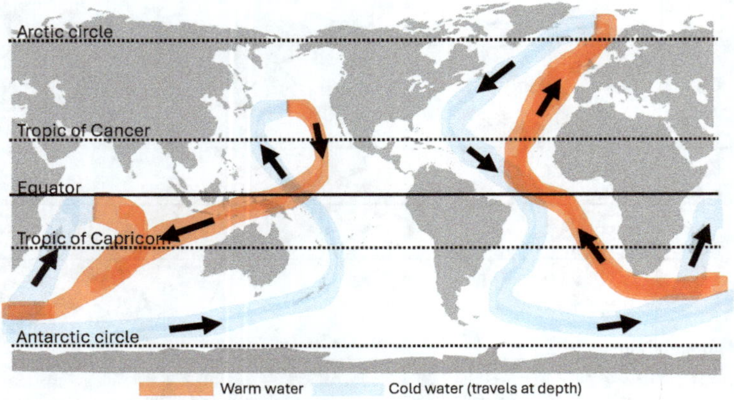

Figure 2.6 Global movement of warm and cold water – cold water travels deep in the oceans.

Source: base map Public Domain, https://commons.wikimedia.org/w/index.php?curid=868126 – the Arctic Circle, Tropic of Cancer, Equator, Tropic of Capricorn, and Antarctic Circle, Antarctic land mass and currents have been added to the base map

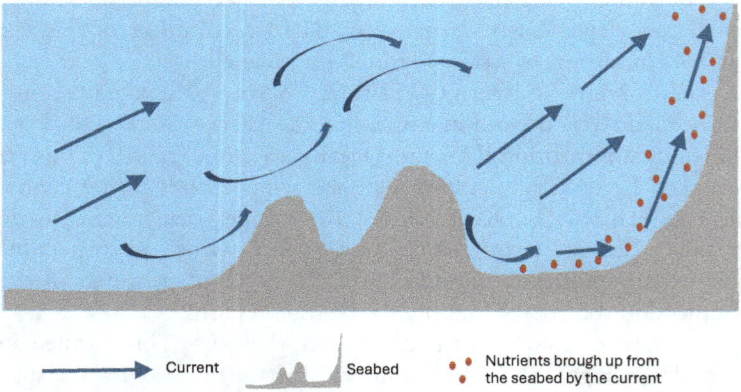

Figure 2.7 Upwelling of nutrient rich, cold water at the coast.

2.3.1.3 Topography

The lithosphere (the ridged, rocky outer layer of the Earth) is characterised by its topography and land-cover. Slopes that face towards the sun will be warmer than those that face away (Figure 2.8). The type of land cover (such as soil, sand, or asphalt) impacts ambient

Figure 2.8 The topography of the Nho Que River, Meo Vac district, Ha Giang Province, Vietnam – the valley creates shade on one side and bright sunshine on the other with some rocks are bare of vegetation because the topography is too steep.

Source: photograph by Gina Hazell

temperature due to heat absorption, reflection, and radiation from warmed surfaces.

As air rises to go over a mountain or mountain range, it cools at a rate of approximately 6.5°C per 1,000 metres up to 11 km (International Civil Aviation Organization 1993). This means that it can carry less water. The water is lost from the atmosphere as precipitation on the windward side (that facing the direction from which the wind is coming) of the mountain or mountain range.

This creates a rain shadow on the leeward side (opposite to the windward) of the mountain or mountain range.

2.3.1.4 Climate diagrams

The climate in any place can be represented in graphical form as a climate diagram – a representation developed by Walter and Leith between 1957 and 1966 (Figures 2.9 and 2.10). Climate diagrams are brief summaries of mean climatic variables throughout a year and are set out in a uniform pattern. Each tic mark along the horizontal axis marks a month. The astronomical summer is placed in the middle of

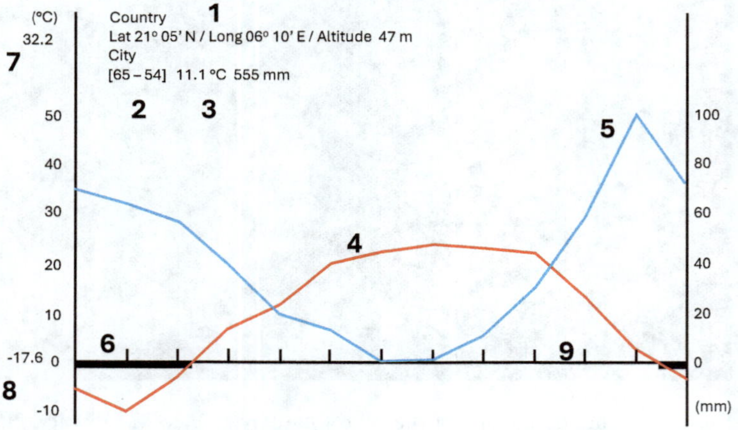

Figure 2.9 Example of a climate diagram. 1: Country name, station location and elevation, station name. 2: Period of observation of temperature (65 years) and precipitation (54 years). 3: Annual average of temperature and annual precipitation total. 4 (red): Temperature average monthly values. 5 (blue): Precipitation average monthly values. 6: Indication of frost periods. 7: Mean daily maximum temperature of the warmest month. 8: Mean daily minimum temperature of the coldest month. 9: The x-axis is the months of the year arranged such that the astronomical summer is placed in the middle of the diagram, hence for places in the Northern Hemisphere January is in the left corner of the diagram and for places in the Southern Hemisphere July occupies this position.

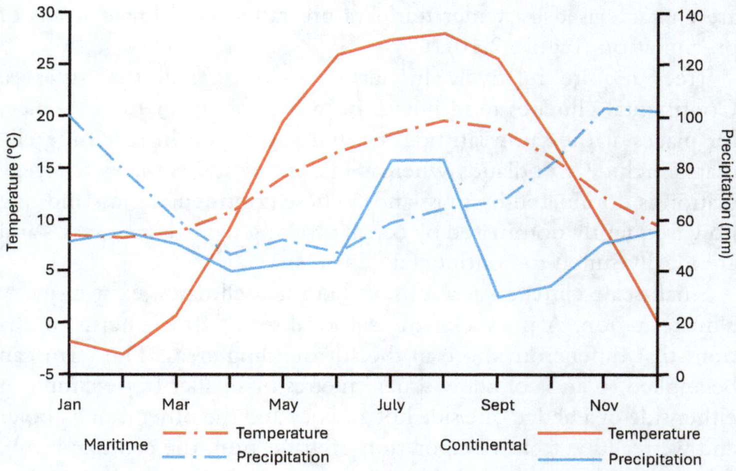

Figure 2.10 Schematic of monthly variation in temperature and precipitation in maritime and continental climates.

the diagram, hence for places in the Northern Hemisphere January is in the left corner of the diagram and for places in the Southern Hemisphere July occupies that position. Monthly precipitation is shown on the right ordinate (y-axis) in 20 mm divisions and mean temperature on the left ordinate in 10°C divisions. When the daily mean minimum temperatures are below zero a black bar is placed immediately below the horizontal line (Figure 2.9).

Ocean and sea currents carry moist air primarily to the west coasts of continents. As this air meets dry land it cools, especially if there are cliffs, hills or mountains that cause the air to rise in altitude, and moisture picked up by the air as it crossed the oceans falls as precipitation from this now cooler air. The result is that precipitation levels are relatively high along west facing coasts. As air flows inland, it dries, and the winter warming effect lessens. Waterbodies tend to gain less heat in the summer than dry land and retain that heat during winter. The effect of this is that maritime climates, near the coast, have a smaller temperature variation (cooler in the summer and warmer in the winter) and have higher precipitation levels, than further inland where continental climates are characterised by wider temperature ranges and lower levels of precipitation (Figure 2.10).

Trees require relatively higher levels of rainfall than grasses. Continental climates tend not to be wet enough for trees. Hence, for places at the same latitude, natural vegetation in maritime climates include woodlands whereas in continental climates the vegetation is dominated by grasslands. These continental grasslands are now frequently dominated by cereal growing (cereals are grasses and are ideally suited to continental climates).

Small-scale climate variations, called microclimates, exist in every climate region. A microclimate is a local set of atmospheric conditions that differ from those in the surrounding areas. The term can be applied to areas of a few square metres or smaller (for example at either side of a hedge one side hot and dry and the other much cooler and wetter (due to less evaporation of water from the plant and soil), the sunny side and shaded sides of an ants' nest) or to areas as large as many square kilometres. Microclimates can be found in most places but are most pronounced in topographically dynamic zones such as mountainous areas, islands, and coastal areas (Ellis and Eaton 2021).

2.3.2 SEASONALITY

Relative to the sun the Earth is tilted at an angle of approximately 23.5°. This means that the angle at which solar radiation hits the Earth's surface varies throughout the year. At the two equinoxes the sun hits the Earth's equatorial surface at 90°. Three months later, at the December and June solstices (i.e. the sun's minimum or maximum declination), when the sun is directly above the Tropic of Capricorn (December) or Cancer (June), solar radiation hits the equator at 66.5°. It is this tilt that causes seasonality: defined as the regular and periodic changes of a condition on an annual timescale. As latitude increases, the angle of incidence of solar radiation decreases, but summer day length (i.e. the duration of sunlight) increases while winter daylength decreases. For example, the angle at which solar radiation hits the Antarctic and Arctic varies from 47° to 0° dependent on the time of year. The effect is that the hours of day light at the Antarctic and Arctic Circle ranges between 24 hours of light and 24 hours of darkness.

Seasonal patterns are influenced by geographic location and climate. Some places have climates with four seasons while others experience three or only two or distinct seasons. Four-season climates (spring, summer, autumn (fall), and winter) are found in temperate zones, including much of North America, Europe, and parts of Asia (Figure 2.11). Three-season climates are common in tropical and subtropical areas. The seasons are typically divided into wet, dry, and cooler periods. India, for example, has a summer (hot and dry), a monsoon season (wet and humid), and a winter (mild and cooler). Similarly, some African countries experience a rainy season, a dry season, and transitional periods. Two-season climates are found in many equatorial regions, including parts of South America, Southeast Asia, and Central Africa. These areas often experience only a dry season and a rainy/wet/monsoon season wet (also known as the 'breeding season', as the associated abundance of resources provides favourable conditions for the survival of offspring).

Seasonality can affect many ecosystem processes and, therefore, the dynamics of populations and communities. Seasonal variables relevant in ecological systems include temperature, photoperiod, rainfall, wind, human activity, upwelling of water from lower levels in the oceans to the surface, and resource (e.g. nutrient) pulses. Infectious disease dynamics may be affected by seasonality through

Figure 2.11 Beech trees (*Fagus sylvatica*) turning colour from green to brown as the season moves from summer to autumn in Abruzzo, Italy.

Source: photograph by Luke Blazejewski

seasonal changes to host behaviour and host immune responses, variations in encounter rates between pathogens and hosts, and seasonal changes in the biology of disease vectors (White and Hastings 2020). Seasonality can promote coexistence between species through temporal niche partitioning (i.e. when two species use the resources available at a given place at different times of the year). This can be seen where, for example, a resident species stores energy (e.g. as body fat) when food is plentiful in the summer and, therefore, does not need to compete with a winter migrant when it is scarcer. Seasonality is also important in community structure (see Chapter 5). In warm, wet periods there is more plant growth and a different structure to a community than in late winter when there has been no growth for many months, plants have died back, and community structures are more open (White and Hastings 2020).

2.3.3 TEMPERATURE AND WATER DEPTH

Water has a high specific heat capacity (i.e. maintains heat and warms or cools slowly), which means that daily temperature fluctuations are only slight and in seasonal latitudes water temperature changes slowly between seasons. As water temperature changes, so does its density. Water, at atmospheric pressure, is most dense at 4°C and as water temperature rises above or fall below this figure, density decreases. In temperate latitudes, the surface of the water warms during the summer. As it warms the water becomes less dense, and floats above and does not mix with colder, denser water beneath. A thermal barrier forms between the warm, less dense and colder, more dense waters. Above this barrier, warmer water, called the epilimnion (from the Greek for 'upper lake') circulates within itself. Below the thermal barrier is the metalimnion (from the Greek for 'middle lake'). This zone is characterised by a steep and rapid decline in temperature by around 1°C for every metre of depth. This abrupt temperature gradient is the thermocline. Deeper still is a cold layer that is cut off from the air by the two layers above it. This is the hypolimnion (from the Greek for 'lower lake'). It is a layer of dense, cold water. The hypolimnion is often low in oxygen and has a layer of bottom mud in which organic matter is decomposed (Figure 2.12).

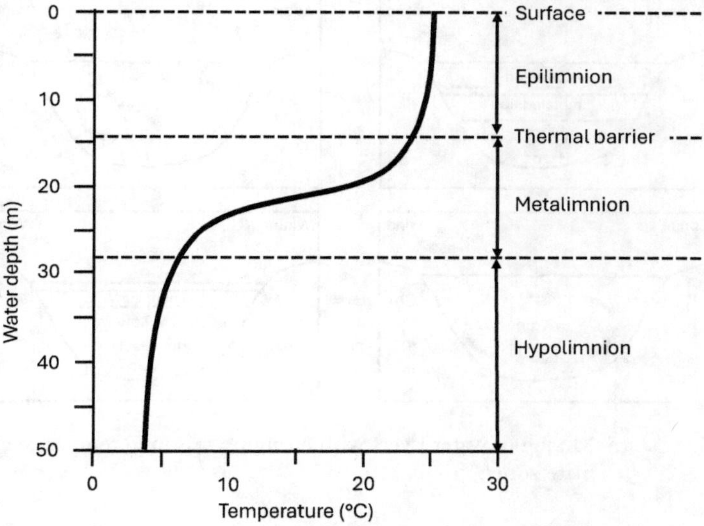

Figure 2.12 Temperature profile with increasing depth of water illustrating the thermocline.

In temperate climates, the air temperature falls during autumn. Once air temperature is below that of the surface water, heat is lost form the water to the atmosphere through convection, conduction, and evaporation. The result is that the water cools. As it does so, the epilimnion expands deeper, the metalimnion sinks and eventually the two layers merge. When the temperature and water density at the top and bottom of the lake are the same, the two layers mix. This seasonal mixing of water is called an 'overturn'. During an overturn, oxygen and nutrients are recharged/refilled throughout the lake. This mixing is further facilitated by winds blowing across the surface of a waterbody (Figure 2.13).

In colder months in temperate climates inverse stratification can occur. This happens when below 4°C floats on the surface of the denser water at 4°C and water temperature rises with increasing depth. This situation can continue until ice forms which reflects most solar radiation. The solar radiation that is not reflected does pass through the ice and warms the water immediately below the

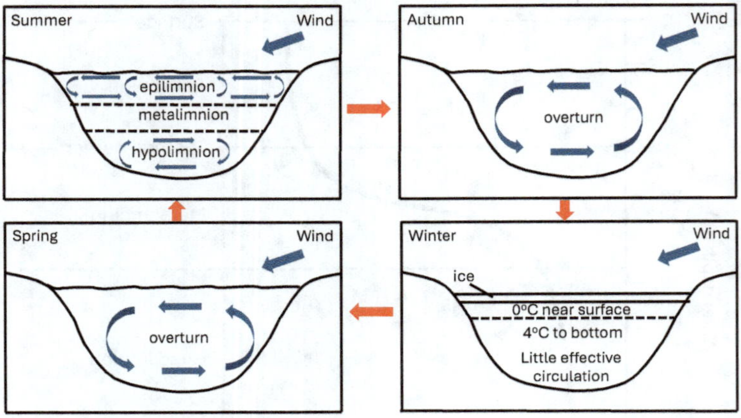

Figure 2.13 Mixing of water layers with changing season in temperate climate zones.

ice. Due to differences in density, this slightly warmer water will sink and mix with that at the bottom of the lake which is warmed by heat from the bottom mud. The result is that the temperature at the bottom of the lake is slightly higher than at the top, though overall thermal stability is maintained.

As the surface ice melts in spring the upper water warms to 4°C and again sinks, aiding further mixing which is enhanced by any wind and results in the whole of the water body mixing. This is the spring overturn. The overturn of water layers brings nutrients sequestered (stored) in the bottom up to the surface and oxygen from the air dissolved in the upper water to the bottom of the lake. Plankton will also be distributed throughout the lake by this water mixing. As spring progresses into summer and air temperatures rise, so the stratification into the epilimnion and metalimnion will become established once more.

This stratification does not occur in ponds and lakes. Temporary, short-term stratification may occur in shallow lakes and ponds and in some shallow water bodies there may be stratification without a thermocline. Further, in very deep lakes (e.g. Lake Malawi, located between Malawi, Mozambique, and Tanzania) the metalimnion may not disappear completely as the epilimnion descends in autumn. In this situation the remnants of the metalimnion and the hypolimnion never mixes with water from higher layers.

2.3.4 PRESSURE (AIR AND WATER)

At sea level the air pressure is around 1.013×10^5 Pa (1 atmosphere). Pressure in the ocean increases by about one atmosphere for every 10 metres of water depth. So, for example, at a depth of 100 metres in salt water, the pressure will be approximately 10.042×10^5 Pa (10 atmospheres) and at 2,000 m below the surface the pressure will be approximately 200.830×10^5 Pa (200 atmospheres). Under the pressures in deep water organisms that have air-filled spaces (such as lungs or a swim bladder as found in mammals and most fish) would be crushed. Organisms that live at depth are largely composed of water, which is incompressible, and, hence, they are not squashed under the high pressure.

In the atmosphere air pressure deceases with altitude. At an altitude of about 5 km, air pressure is roughly half and at 10 km roughly one third of that at sea level (the top of Mount Everest is approximately 8.8 km above sea level). The thin air at altitude contains low overall amounts of oxygen, meaning that many species cannot survive. The low air pressure is also a determining factor in how high some birds can fly as the thin air provides reduced lift compared to more dense air.

2.3.5 LIGHT

Light is only able to penetrate a certain distance into a water column. This is due to the cloudiness of water caused by suspended particles and to light being intercepted by algae and plants growing on or near the surface of the water. In very clear conditions sunlight may travel down about 1,000 m, however, there is rarely any significant light, and hence no photosynthesis, beyond 200 m. In the terrestrial environment the main issues regarding light are the duration of daylight (which is linked to seasonality) and ultraviolet radiation. The 23° tilt of the Earth relative to the Sun means that the duration of the light (i.e. daylength) varies through the year. The extent of the variation is related to latitude (see section 2.3.2). With the advent of artificial sources of light and the installation of street lighting and lighting in homes and commercial premises, there is an increase in artificial light at night (ALAN). This is an issue associated with urbanisation and is discussed in detail in Chapter 3.

Sunlight contains a small amount of short wavelength ultraviolet (UV) light irradiation which is harmful to life. Most of the UV irradiation is filtered out by the stratospheric ozone layer (between 15 and 30 km above sea level). Ultraviolet radiation is damaging to plant and animal tissues. Higher than normal levels of UV (especially UV-B) damages plants. Examples include damage to the DNA and to membranes, and photosynthetic or hormone systems disorders (Piri et al. 2019). Annual and deciduous plants lose the UV damaged leaves during their dormant period and grow new leaves when they become active again.

Different groups of animals perceive different wavelengths of light. Many mammals have colour vision to a greater of lesser extent (some are colourblind (e.g. dogs (*Canis* spp.)), seeing their environment only in shades of grey). Birds and some insects can see ultraviolet light in addition to the red to blue wavelength light visible to humans. Taken together with differences in auditory range (see next section), olfactory and other specialised senses (e.g. the ability to detect magnetitic fields, see polarised light, detect electric fields) means that different groups of animals perceive the environment around them in very different ways. The word 'unwelt' is used to mean the world as it is experienced by a particular animal.

2.3.6 NOISE

The natural world is full of sound: geophony, biophony, and anthrophony (Table 2.4).

Many animals communicate through sound. For successful communication, their calls and songs must be heard and be recognisable. While most mammals detect sounds between 20 Hz and 20 kHz, there are many animals that hear sounds beyond this spectrum.

Table 2.4 Sound sources

Source	*Examples*
Geophony – sounds made by the physical environment	The sound of the wind, waves breaking on a shore, a waterfall, rustling of leaves, and thunder
Biophony – sounds made by animals	Bird song, contact calls between animals, and stridulation (the making of a shrill sound by rubbing the legs, wings, or other parts of the body together) by insects
Anthrophony – sounds made by people	Talking, singing, music, traffic noise, noise from industrial plants, aircraft noise, and noise from ships and boats including sonar

Dogs (*Canis* spp.), for example, perceive frequencies up to 45 kHz, and during echolocation bats (Chiroptera) use sounds above 100 kHz. In the marine environment, species such as dolphins detect ultrasonic pulses up to 150 kHz. At the other end of the frequency spectrum, elephants (Elephantidae) are sensitive to infrasonic frequencies below 20 Hz. Noise generated by people is becoming more omnipresent in the natural world and is interfering with animal communication. There will be more on this topic in the case study – urbanisation in Chapter 3.

2.4 CASE STUDY – ECOLOGICAL IMPLICATIONS OF A CHANGING CLIMATE

Climate change drives change in both the natural and social environment (Figure 2.14).

The ecological implications of climate change include changes in the abundance, location, and functions of organisms (their interactions and their effects on the abiotic environment) (Table 2.5).

Some generalities of the impacts of climate change were drawn together by Rickwood et al. (2025). A significant headline statement in their paper is that climate change increases extinction risk of many species. One of the key drivers of this increase is a change in the phenology – the seasonal and interannual timing of biological events – of plants and animals.

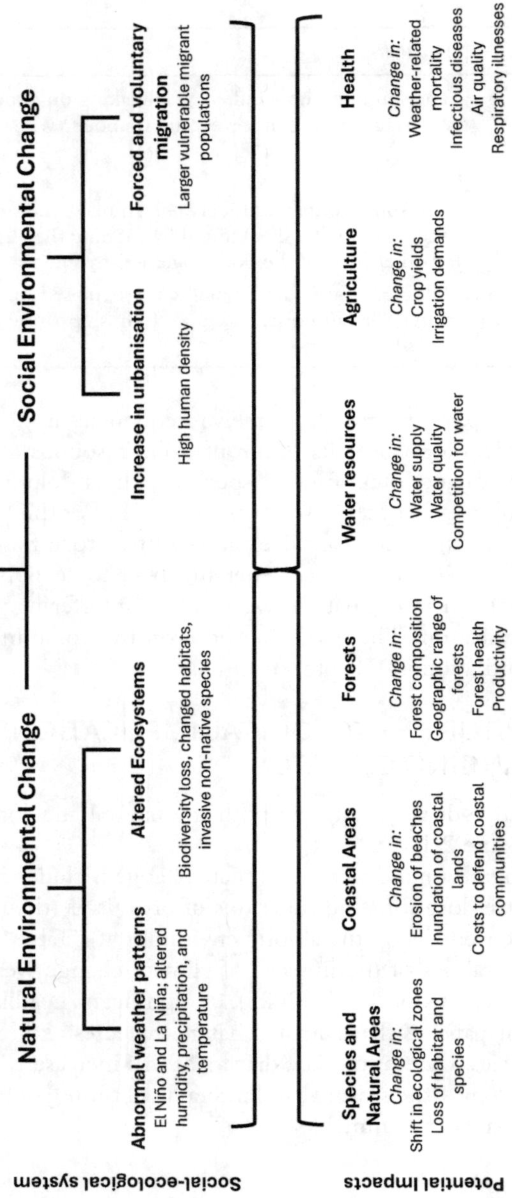

Figure 2.14 The impacts of climate change on abiotic and biotic aspects of the environment.

Source: based on Williams et al. (2023), with additions

Table 2.5 Some ecological implications of a changing climate

Changes to	*Examples*
Abiotic factors (this chapter)	Increased erosion and loss of habitats due to rising sea levels and increased storm intensity
Species (see Chapter 3)	Changes in phenology, morphology, genetic frequency, physiological, behavioural, and life history changes
Populations (see Chapter 4)	Shifting ranges of species poleward or up in elevation Changes in population density
Ecosystems (see Chapter 6)	Changes to trophic structures Reduced productivity in agroecosystems and silvicultural systems
Landscapes (see Chapter 7)	Decreased connectivity Changing landscapes

Source: based on Malanson and Alftine (2023)

Migratory species time arrival at their breeding grounds to optimise nest site choice, increase mating opportunity, maximise food abundance, or enable multiple or replacement clutches. Several factors influence the timing of arrival including the phenology and abundance of food sources at overwintering grounds and stop-over sites, carry-over effects of environmental conditions at overwintering grounds and along migratory routes, the cueing of behaviours by non-environmental factors such as day length (i.e. photoperiod), and differences in the rate of environmental change between overwintering and breeding grounds. Climate change can result in asynchrony between the timing of arrival at the breeding grounds and the optimal conditions for reproduction, which may lead to reduced reproductive output, population declines and biodiversity loss.

The issue is, then, how can ecological knowledge be applied to address the ecological impacts of climate change? Some interventions are set out in Figure 2.15. Throughout this volume, information will be set out in the case studies which present greater detail on some of these interventions.

In this case study, the focus has been on a sub-set of climate related issues – the ecological implications of climate change. This is not to dismiss the other issues, and those interested in finding out more are encouraged to do so.

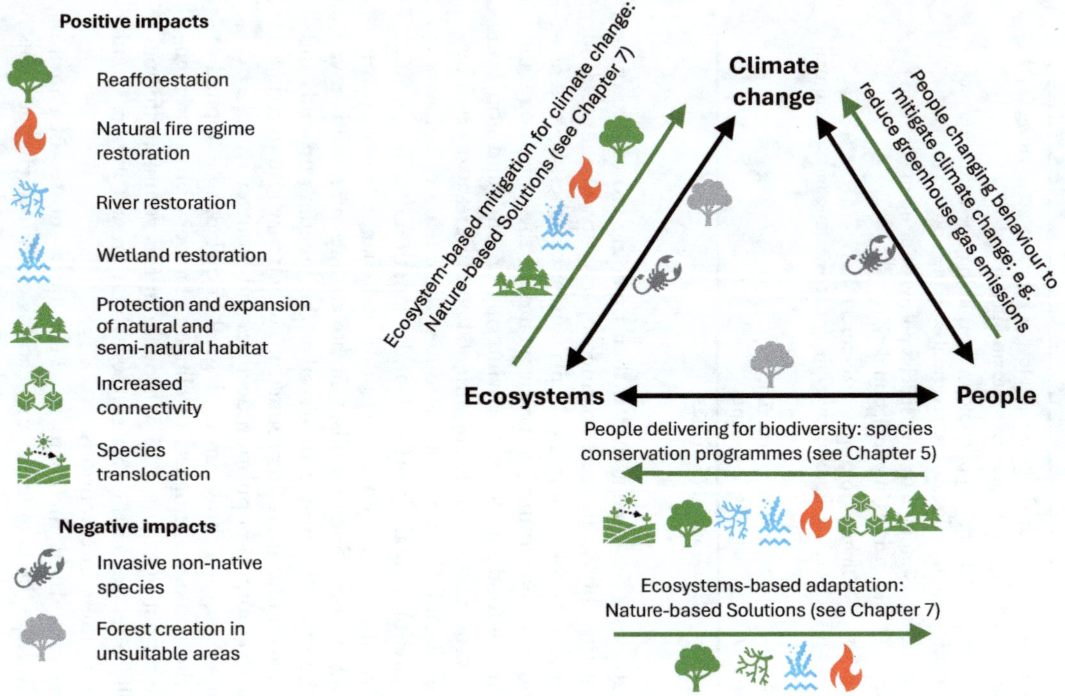

Figure 2.15 The impacts of climate change and possible mitigation strategies.

Source: based on Morecroft et al. (2019)

2.5 SELF-STUDY SUGGESTIONS

2.5.1 SELF-STUDY QUESTIONS

1. What are the meanings of biotic and abiotic?
2. What are the main abiotic factors in the environment?
3. How does the concentration of salt in water affect the animals that live in that water?
4. How do marine animals confront the challenges of thermoregulation and osmoregulation?
5. How far does light penetrate into water and what are the consequences of this?
6. How does temperature vary with water depth and what are the consequences?
7. What chemical reaction takes place when carbon dioxide dissolves in water?
8. What levels of carbon dioxide in water are stressful for fish?
9. What are the differences you would expect to see between lakes that are oligotrophic, mesotrophic, and eutrophic?
10. What is soil?
11. What are the main ecological functions of soil?
12. What are the differences between acidic, neutral, and alkaline (or base-rich) soils?
13. What is the difference between climate and weather?
14. What are the effects of seasonality?
15. Why are atmospheric cells important in terms of global climate patterns?
16. What are the differences between a maritime and continental climate and how do ocean currents contribute to these differences?
17. What is the general pattern of vegetation observed in moving from the equator to the poles?
18. Why are microclimates important?
19. How does climate influence the distribution and structure of Earth's terrestrial biomes?
20. What factors are most important in determining the distribution and structure of the Earth's aquatic biomes?
21. How might a changing environment affect the global distribution of biomes?

2.5.2 A DEEPER DIVE

A. ClimateCharts.net is a web-based resource in which it is possible to view climate diagrams for locations across the globe. Select at least ten sites on ClimateCharts.net and compare the climate charts, compare sites that are close to an ocean and those that are in the centre of a continent. Compare sites that are near the equator with those that are in temperate and cold parts of the globe. Compare sites that are in deserts with those in grassland and forested areas.
B. Search the internet for 'Soil Map of the World' and explore that site.
C. Visit https://gml.noaa.gov/ccgg/trend/graph.html to view data on the level of atmospheric carbon dioxide. How has the level of atmospheric carbon dioxide changed in the last 60 years? By how much has it changed? From this page it is also possible to view tends for atmospheric methane (CH_4), nitrous oxide (N_2O), and sulphur hexafluoride (SF_6). How has the level of these molecules in the atmosphere changed over time? Why are these levels of these molecules important?

2.6 REFERENCES

Anon (2025). 'Comparing mineral content in different soil types'. Retrieved from https://livetoplant.com/comparing-mineral-content-in-different-soil-types (accessed 1 August 2025).

Brown, G.E. Jr. and Calas, G. (2011). 'Environmental mineralogy – understanding element behavior in ecosystems'. *Comptes Rendus Geoscience* 343(2–3): 90–112. https://doi.org/10.1016/j.crte.2010.12.005.

Ellis, C.J. and Eaton, S. (2021). 'Microclimates hold the key to spatial forest planning under climate change: cyanolichens in temperate rainforest'. *Global Change Biology* 27(9): 1915–1926. https://doi.org/10.1111/gcb.15514.

Gough, W.A. and Leung, A.C.W. (2022). 'Do airports have their own climate?'. *Meteorology* 1(2): 171–182. https://doi.org/10.3390/meteorology1020012.

International Civil Aviation Organization. (1993). *Manual of the ICAO Standard Atmosphere (Extended to 80 Kilometres (262 500 Feet))* (3rd edition). Montréal: International Civil Aviation Organization.

Malanson, G.P. and Alftine, K.J. (2023). 'Ecological impacts of climate change'. In R. Sivanpillai and J.F. Schroder (eds), *Biological and Environmental Hazards, Risks, and Disasters* (2nd edition). Amsterdam: Elsevier. https://doi.org/10.1016/B978-0-12-820509-9.00021-6.

McCabe, K. (2022). 'Jet stream and stormy weather'. Retrieved from www.rmets.org/metmatters/jet-stream-and-stormy-weather.

Morecroft, M.D., Duffield, S., Harley, M., Pearce-Higgins, J.W., Stevens, N., Watts, O. and Whitaker, J. (2019). 'Measuring the success of climate change adaptation and mitigation in terrestrial ecosystems'. *Science* 366: 6471. https://doi.org/10.1126/science.aaw9256.

Piri, E., Babaeian, M., Tavassoli, A. and Esmaeilian Y. (2019). 'Effects of UV irradiation on plants'. *Advanced Journal of Microbiology Research* 13(3): 001–007. https://doi.org/10.5897/AJMR11.263.

Rickwood, M.L., Tucker, E., Beton, D., Davey, S., Godley, B.J., Snape, R.T.E., Postma, E. and Broderick, A.C. (2025). 'Individual plasticity in response to rising sea temperatures contributes to an advancement in green turtle nesting phenology'. *Proceedings of the Royal Society B* 292: 2041. https://doi.org/10.1098/rspb.2024.1809.

US EPA. (2023). 'Indicators: dissolved oxygen'. Retrieved from www.epa.gov/national-aquatic-resource-surveys/indicators-dissolved-oxygen (accessed 15 February 2025).

White, E.R. and Hastings, A. (2020). 'Seasonality in ecology: progress and prospects in theory'. *Ecological Complexity* 44: 100867. https://doi.org/10.1016/j.ecocom.2020.100867.

Williams, P.C.M., Beardsley, J., Isaacs, D. Preisz, A. and Marais B.J. (2023). 'The impact of climate change and biodiversity loss on the health of children: an ethical perspective'. *Frontiers in Public Health* 10: 1048317. https://doi.org/10.3389/fpubh.2022.1048317.

3

INDIVIDUALS

Adapting to the environment

3.1 INTRODUCTION

Abiotic factors (see Chapter 2) vary across the global. A framework for understanding how environmental abiotic factors influence an organism's ability to survive is provided by a series of 'laws': the law of the minimum, the law of limiting factors, and the law of tolerance. These laws are discussed in section 3.2. Together these laws help an ecologist understand the global and local distribution of plants and animals, and how natural and human caused environmental change affects these distributions.

To survive an individual must be adapted to the environment in which it lives. These adaptations may be morphological, physiological, and/or behavioural. In section 3.3 the focus is on the adaptations shown by plants and animals living under different environmental conditions. The environmental factors considered are temperature, light (including seasonality), water levels (including drought and flooding), salinity, nutrient type and availability, and climate (see Chapter 2 for descriptions of these environmental factors). Understanding these adaptations and how they relate to the environment provides insights that help to shape conservation strategies, appreciate the consequences of a changing environment, and inform practices in agriculture, forestry and other fields where an understanding of plant and animal behaviour and physiology are important.

Globally, urban areas are growing rapidly both in terms of their number and size. Urban areas are not, as was once thought, devoid

DOI: 10.4324/9781003604464-3

of wildlife, rather there are many species adapted to live and thrive in the unique urban environment. Urban areas present plants and animals with new environmental challenges including some not encountered in the natural environment. While some species cannot tolerate these conditions, others thrive. Building from the theories discussed in this chapter, the case study focusses on how plants and animals adapt to urbanisation.

3.2 LIMITING FACTORS AND TOLERANCE

3.2.1 THE LAW OF THE MINIMUM AND THE LAW OF LIMITING FACTORS

In 1840, Justus Leibig (aka Justus Freiherr von Liebig, 1803–1873, a German organic chemist) published *Organic Chemistry and Its Applications to Agriculture and Physiology*. In this book, Liebig set out his conclusion that plant growth responds to the nutrient that is most limiting. This statement became known as the law of the minimum (aka Liebig's law of the minimum), which states that growth of an individual is not dictated by the total amount of resources available, but by the single factor in shortest supply. For example, plant growth may be limited by a lack of available nitrogen in the soil, even if light, water, and other nutrients are abundant. In this example, the amount of nitrogen is the limiting factor. More generally, a limiting factor may be chemical, physical, or biological (e.g. the presence of a predator or disease-causing organism). Strictly, this law only applies under certain conditions: when all other resources needed by the plant are present in amounts that are more than the plant needs and remain in a steady state. Under these conditions, if the amount of the limiting factor is increased, the growth of the plant increases. If the amount of that limiting factor continues to increase there will come a point when it is no longer the limiting factor, and plant growth will be limited by the availability of some other resource. The principle of the limiting factor determines the carrying capacity of an environment (i.e. the maximum population size of a species that an environment can sustain indefinitely, given the available resources such as food, water, and habitat) and restricts species to specific geographical regions.

Sixty-five years later, in 1905, F.F. Blackman (1866–1947, a British plant pathologist) noted that plant growth is also restricted if the

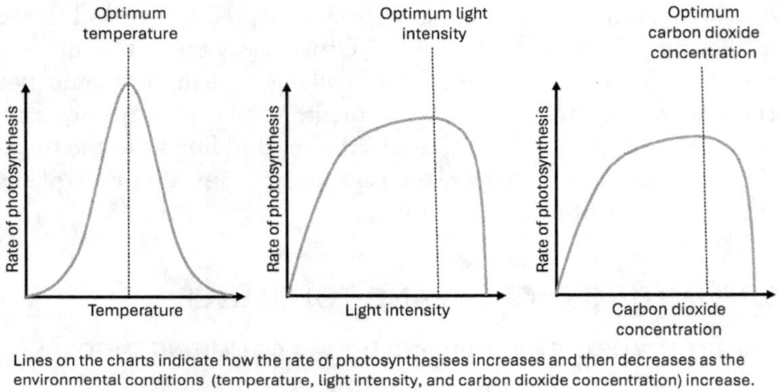

Figure 3.1 Relationships between temperature, carbon dioxide concentration, and light intensity with rate of photosynthesis.

availability of a resource exceeds a certain maximum. Blackman's observations became known as the law of limiting factors. Take, for example, the factors that affect the rate of photosynthesis in a plant leaf. The relationships between the rate of photosynthesis and three main factors (light intensity, carbon dioxide concentration, and temperature) are shown in Figure 3.1. The curves in Figure 3.1 illustrate that there is an optimum, species specific, temperature, light intensity, and carbon dioxide concentration for photosynthesis; below that optimum biochemical processes do not operate as quickly as they might and above that optimum enzymes are deactivated or destroyed, and photosynthesis slows and eventually stops.

Blackman's principle of limiting factors states that when a process is governed by more than one factor, the rate of that process is governed by the factor which is closest to its minimum or maximum value (Figure 3.2).

3.2.2 THE LAW OF TOLERANCE

Both the laws of the minimum and of limiting factors emphasise environmental conditions. In 1913, Victor Ernest Shelford (1877–1968, an American zoologist and animal ecologist) turned the idea around, reframing it in terms of the organism and what an organism can tolerate. This became known as the law of tolerance

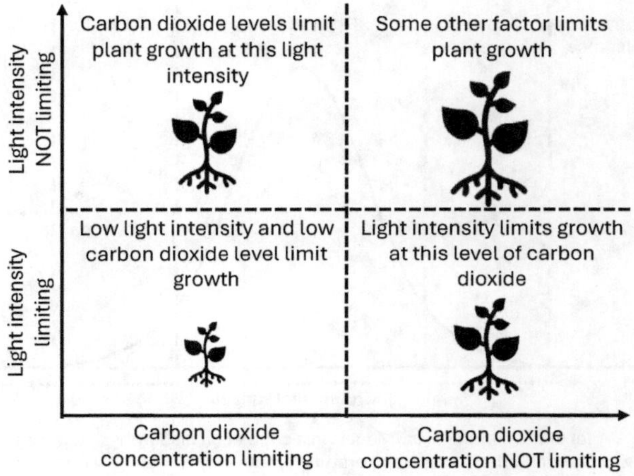

Figure 3.2 Blackman's principle of limiting factors using the rate of photosynthesis and plant growth as an example.

(aka Shelford's law of tolerance) which states that an organism's success is based on a complex set of conditions and that each organism has a certain minimum, maximum, and optimum environmental factor or combination of factors that determine success. It is often explained using a diagram (Figure 3.3).

The width of the range of tolerance is species and environmental resource specific. Outside a species' optimal space, in the zone of stress, the organism will survive but will not grow as rapidly or to the same size and will not have the same reproductive success as under the optimum conditions. Even further from the optimum conditions is the zone of intolerance. In this zone the organism cannot survive and dies or, if it is able, it moves to an area in which the environmental conditions are more optimal.

Figure 3.3 illustrates Shelford's law for one environmental factor; however, multiple factors operate at any given time. According to Shelford's law of tolerance, the presence, abundance, and distribution of a species are determined by whether the levels of the multiple physical and chemical factors fall within a specific range the organism can withstand. This would mean drawing Figure 3.3 in *n* dimensions, which is not possible. This means that it is left to

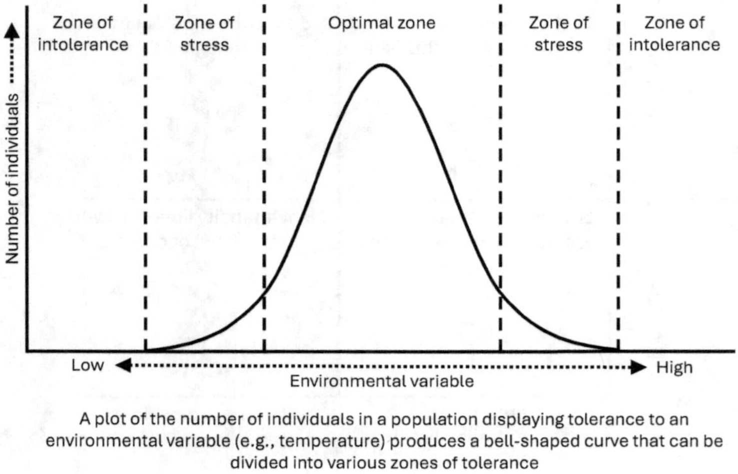

Figure 3.3 Shelford's law of tolerance.

the ecologist to imagine this *n*-dimensional space in which every species has an optimal zone where it performs best, surrounded by stress zones where survival is taxing, and finally, maximum and minimum limits beyond which it cannot persist. Organisms with a wide range of tolerance for multiple abiotic factors are the most widely distributed species. Conversely, slight shifts in the abiotic environment, such as increased water salinity or a sustained rise in average temperature, can result in sensitive species being in an environment that that is outside their tolerance range. This forces populations to either migrate or face local extinction, fundamentally altering the biodiversity and balance of the ecosystem.

Tolerance ranges are not necessarily fixed. Variations in tolerance between species and individuals of the same species arise from differences in basic biochemistry, physiology, morphology, genetics, and behaviour. No single set of characteristics enables an organism to function equally well under all environmental conditions. As environmental conditions change, so do the set of characteristics needed to survive. Changes in tolerance can also be associated with the life stage of the organism. For example, the nymphal stages of some invertebrates (including dragonflies and damselflies (Odonata), stoneflies (Plecoptera), and mayflies (Ephemeroptera)) are aquatic while the adults are terrestrial. Seasonality (see Chapter 2) can also require resident species to change their tolerance of environmental conditions. For example, fish living in a pond are adapted to live in cold water in the winter and warmer water in the summer; they cannot tolerate being moved suddenly from water at winter temperatures to water at summer temperatures and *vice versa*. During spring and autumn, fish adapt to the gradually changing water temperatures. This adaptive process is known as acclimatisation and is found widely in plants and animals that live in temperate environments. Another example of acclimatisation is the seasonal growth and shedding of coats by many animals.

3.3 PLANT AND ANIMAL ADAPTATIONS TO DIFFERENT ENVIRONMENTAL CONDITIONS

Plants face many environmental challenges including tough environmental conditions, predation by herbivores, and competition from other plants for resources including sunlight and water. The main adaptions seen in plants are structural (morphological), physiological, reproductive, and behavioural (Figure 3.4). An individual plant species is likely to have many adaptations across all four major categories.

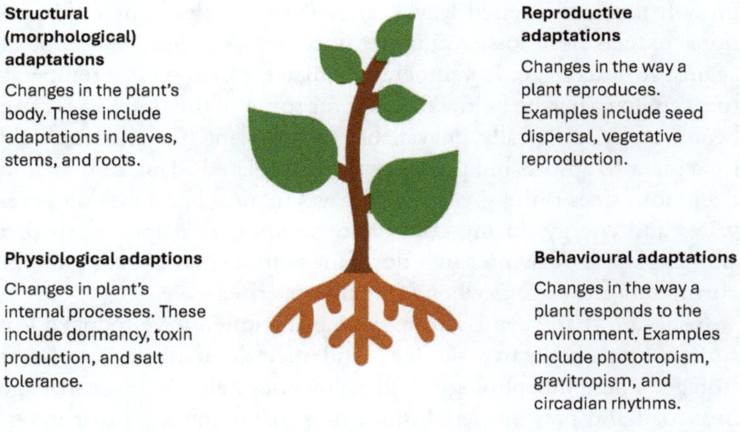

Figure 3.4 Adaptations of plants to their environment.

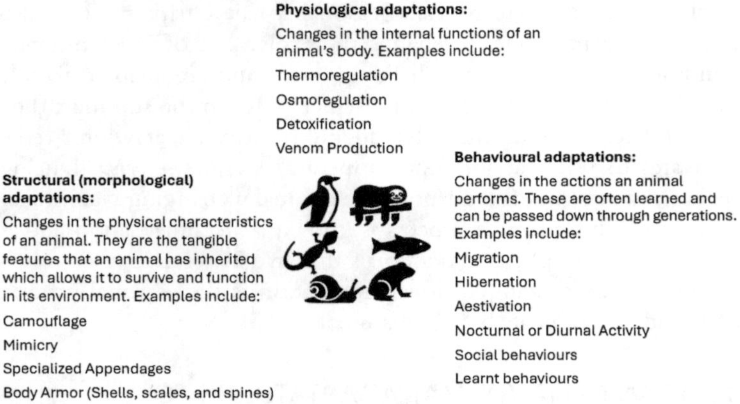

Figure 3.5 Adaptations of animals to their environment.

In the same way that plants have different adaptations, so do animals (Figure 3.5), and, as with plants, an individual animal species is likely to use a range of adaptations.

3.3.1 ADAPTATIONS TO TEMPERATURE

Plants show range of adaptations to temperature. In colder climates, many plants have morphologically adaptions including a compact growth form, thickened leaves, and waxy cuticles. These adaptations reduce heat loss, minimise water evaporation, and protect against frost damage. In temperate climates where winter temperatures are low (maybe below freezing for some of the time) water can become physiologically unavailable to the plant (i.e. it is too cold for a plant to absorb and process water). A related adaptation seen in deciduous trees is the shedding of leaves in autumn. This conserves water and energy during cold winters. Another adaptation is that annual plants overwinter in a dormant state (e.g. as a seed, bulb, or corm) and grow again when temperatures rise.

In hot, arid and semi-arid regions plants often have reduced leaf surface areas, reflective surfaces, and specialised tissues for water storage. These morphological characteristics help the plant manage heat load and prevent desiccation, thus maintaining cellular integrity. For example, cacti have thick, fleshy stems that store water and perform photosynthesis, spines (modified leaves) which reduce water loss (and also deter herbivores), a waxy cuticle that reflects heat and reduces moisture loss, and shallow, widespread roots that can seek out water and absorb water when there is any precipitation.

Plants also have numerous physiological and molecular adaptations to temperature. For example, in cold environments, some plants produce antifreeze proteins which inhibit ice crystal formation within tissues and reduce the risk of frost injury. In hot conditions some species synthesise heat shock proteins which function as molecular chaperones to repair and stabilise denatured proteins. Another example is the alteration in the lipid composition of cellular membranes. Such alterations ensure that membrane fluidity is maintained across temperature swings – a process essential for proper cellular functioning.

Green plants fix atmospheric carbon dioxide into organic carbon molecules using solar energy. This biochemical process is called photosynthesis (see Chapter 6 for a comparison of photosynthesis and chemosynthesis). Plants have adapted to different environmental conditions by developing three different photosynthetic routes known as C_3, C_4, and CAM (crassulacean acid metabolism). The nomenclature refers to differences in the intermediate chemicals in photosynthesis. C_3 plants are adapted to temperate climates with ideal temperatures 20–30°C. C_4 plants are adapted to hotter climates with ideal temperatures 30–40°C. Most C_4 plants are grasses native to tropical and subtropical regions and some shrubs and dicotyledonous herbaceous plants found in arid or saline environments. CAM plants are adapted to desert climates with hot days around 30–40°C and cold nights with temperatures below freezing around −5°C.

Animals adapt to different temperatures by changes in physiology and/or behaviour. Physiologically, animals can be divided into three groups. Poikilotherms allow their internal body temperature to vary with the external temperature (e.g. invertebrates, fish, amphibians, and reptiles). Homeotherms maintain a fairly constant internal temperature irrespective of the external temperature (e.g. birds and most mammals). Heterotherms sometimes regulate their internal temperature and sometimes do not (e.g. bees (Apodidea) and bats (Chiroptera)).

Poikilotherms use external heat sources (i.e. ectothermy) such as solar radiation to establish an internal temperature at which their metabolism can work sufficiently to allow them to move and digest food. One such example is the marine iguana (*Amblyrhynchus cristatus*) found on the Galápagos Islands which are in the Pacific Ocean 965 kilometres off the coast of Ecuador. The ocean around the Galápagos Islands ranges from 22°C to 28°C which is cooler than the temperature at which vertebrate metabolism is most efficient. Before diving to find food, marine iguanas bask on rocks orientating themselves with respect to the sun to warm up their bodies. While diving they lose heat and their metabolism slows such that when they emerge onto dry land, their movements can be very sluggish. Out of the water, they bask once more to warm up, raising their metabolic rate, before diving again. They are also known to sleep in piles, one animal on top of one another. This behaviour is an example of communal thermoregulation – where a group of animals cooperate to regulate their body temperature. By huddling together in large groups, the marine iguanas are less exposed to the cool night air and by sharing their body heat they maintain their body temperature.

Being poikilothermic offers both advantages and disadvantages. As poikilotherms do not depend on internally generated heat, they can have a relatively low metabolism and, hence, can withstand relatively long periods of food and water shortage. They are also not restricted by limitations placed on body shape and size that are associated with being homothermic. This allows for a range of lifestyles, for example, earthworms (Annelida), millipedes (Diplopoda), centipedes (Chilipoda), snakes (Serpentes), and single celled organisms such as paramecia (single-celled protozoa that thrive in water). In general, ectotherms live longer than homeotherms. Conversely the disadvantages include being able to be active only when ambient temperature is at the appropriate level or when basking can raise internal the body temperature sufficiently. Hence, poikilotherms can be sluggish in mornings before they warm up sufficiently, and again in the evenings as they cool.

Homeotherms maintain their internal temperature by oxidising glucose and other energy-rich molecules, hence, they can remain active regardless of the ambient temperature. Smaller animals lose relatively more energy to the environment as they have a greater surface area to volume ratio than larger animals. This high energy cost requires small mammals to consume food frequently, for example shrews (Soricidae) need to eat 200 to 300 per cent of their body weight in food each day and this means that they must feed at least every 2 to 3 hours (Churchfield et al. 2012). Within any taxonomic group, smaller animals have a higher basal metabolic rate than larger animals. Basal metabolic rate is the amount of energy an animal requires while resting. As basal metabolic rate is proportional to body mass, this puts an upper limit on how big homeotherms can become. Conversely as body mass decreases the basal metabolic rate increases and with very small body sizes the increase is exponential. This puts a limit to how small a homeotherm can be. The smallest homeothermic mammal is the white-toothed shrew (*Suncus etruscus*) which weighs around 2.4 grams and has a basal metabolic rate of 6.00 ml of O_2 hr^{-1} g^{-1} (ml of O_2 hr^{-1} g^{-1} means millilitres of oxygen per hour per gram [of the animals' weight]) equivalent to 14.4 ml of O_2 hr^{-1}. (ml of O_2 hr^{-1} means millilitres of oxygen per hour). This rate can be compared with the much slower rate of a western moose (*Alces alces*) which weighs 325 kg and has a basal metabolic rate of 0.16 ml of O_2 hr^{-1} g^{-1}. Of course, as the western moose is much larger than the white-toothed shrew, the moose consumes 51,419 ml of O_2 hr^{-1} (White and Seymour 2003).

Homeotherms utilise several physiological and morphological mechanisms to regulate their temperature. One example is counter-current heat exchange (Figure 3.6). This mechanism can be found in the legs of birds, the flippers of dolphins (Odontoceti), the tail of the beaver (*Castor* spp.), and the legs of some mammals (e.g. Arctic fox (*Alopex lagopus*), wolf (*Canis lupus*), and sled dogs, where the footpads of these animals are routinely around 0°C (Hill 1976). In the example in Figure 3.6, arteries carry warm blood at around 35°C from the animal's body into its legs. These arteries are in close contact with the veins that carry blood from the leg back into the body. Heat is transferred from the blood in the artery to that in the vein. The effect is that as blood travels further into the limb, it becomes cooler, and, therefore, loses less heat to the environment. At the same time, venal blood flows back towards the body core and as it does, it warms so that the body is not cooled by cold/cool blood entering from the legs.

In high temperatures, homeotherms must cool down. They may do this by morphological, behavioural, and/or physiological adaptations. Morphological adaptions allow certain parts of the body to

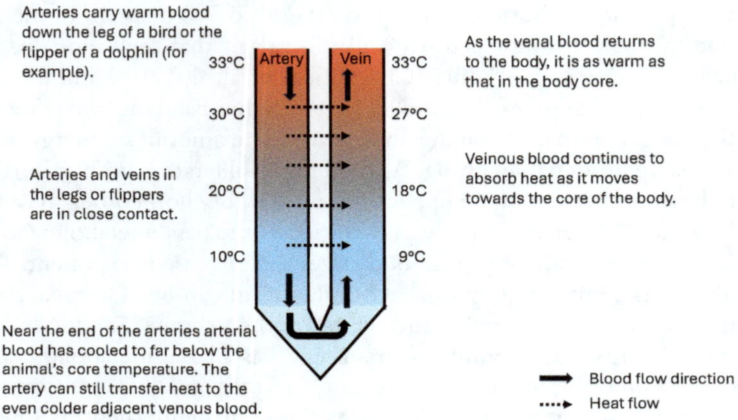

Figure 3.6 Counter-current heat exchange.

act as thermal windows which radiate heat, for example, the large ears (Schmidt-Nielsen 1964). In 1877, Joel Asaph Allen (1838–1921, an American naturalist) stated that the body surface area to volume ratio for homeothermic animals varies with the mean temperature of the habitat to which they are adapted (i.e. the ratio is low in cold climates and high in hot climates). In general terms this means that animals adapted to cold climates have shorter and thicker limbs and body appendages than animals adapted to warm climates. This has become known as Allen's rule (Figure 3.7). As with all generalisations, there are exceptions, but the rule is seen to apply in many cases. It has also been noted that some animals are shifting their morphology to have proportionately larger appendages in response to climate change and the associated temperature increases (Ryding et al. 2021).

Homeothermic animals living in hot and arid climates have developed a range of behavioural strategies. One common adaptation is nocturnal activity, where species such as desert fox (*Vulpes zerda*) and many rodents avoid the daytime heat by being active at night. This minimises exposure to intense sunlight and reduces water loss through evaporation. During the day these animals seek shade and/or burrow. Some kangaroos (Macropodidae) and rabbits (Lagomorpha) lick their fur to promote evaporative cooling, while certain insects extract moisture from food sources to stay hydrated.

Figure 3.7 Two species showing features of Allen's rule. (a) Adult male polar bear (*Ursus maritimus*) with relatively short, thick limbs and small ears adapted to the cold climate off the coast of Spitsbergen, Norway. (b) African wild dog (*Lycaon pictus*) with relatively long, slender legs and large ears adapted to a hot environment of the Lower Zambezi, Zambia.

Sources: (A) photograph by Brian Roger; (B) photograph by Luke Blazejewski

Physiological adaptations to cope with high temperatures include panting and sweating. Both adaptations allow evaporative cooling. Evaporative cooling does come at the cost of losing water from the body, which will then need to be replaced. Panting is very rapid shallow breathing that increases evaporation from the upper respiratory tract. It is an energy-using activity. A variation on panting used by some birds is gular fluttering. This is the rapid vibration of the gular area (a highly vascular oral membrane) while keeping the mouth open. Sweating, excreting moisture through specialised skin glands, occurs only in some mammals, and then not necessarily across the entire body (e.g. fox, wolf, and domesticated dog (all Canidae) only have moisture releasing sweat glands on their paw pads). Some animals in hot environments have one or more discrete vascular bundle(s) of many small, parallel, intermingling vessels called a rete (plural – retia). As with the counter-current arrangement (see Figure 3.6), heat exchange takes place in a rete (e.g. there are retia in the heads of the large antelopes (*Oryx* spp.) that help the animal loose heat and keep its brain cool).

Homeothermic animals must also tolerate cool and cold conditions. Birds and mammals can maintain their core body temperature by changing the thickness of insulating hair, fur, feathers, and fat. Prior to winter, many mammals feed up, laying down fat to act as an energy reserve. Many mammals also acquire a heavier coat while birds add to their thermal stability by fluffing their feathers in cold temperatures (to trap air). If the temperature falls below that which the insulating layers can compensate homeothermic animals increase their metabolism to generate more heat.

When the body temperature of birds and mammals falls below a critical level, shivering occurs. This is the uncoordinated, involuntary, high-frequency contraction of skeletal muscle, which converts chemical energy into heat energy. Non-shivering thermogenesis can also occur. This is when heat is generated from the metabolism of brown fat, a highly vascular brown adipose tissue found around the head, neck, thorax, and major blood vessels. The heat generated allows mammals to be active in temperatures where shivering produces insufficient heat.

In cold conditions, birds and mammals may huddle together to conserve heat. For example, wrens (*Troglodytes troglodytes*) have been found huddled together in nesting boxes and other crevices in cold conditions. Similarly, Japanese macaques (*Macaca fuscata*) huddle in groups and use onsens (hot springs) to keep warm and meerkats (*Suricata suricatta*) huddle together underground to maintain a stable temperature during the cold nights.

Seasonal migration is a strategy used by many species to avoid cold conditions. Birds follow fly ways from areas where winter temperatures are low and, as a result, food supply is low or not available as it is frozen or covered in snow to places that are warmer and where food is available (Figure 3.8). A different migratory pattern is seen in the Antarctic ocean, where zooplankton (1–10 mm in size) migrate from the surface where they eat phytoplankton in the relatively warmer summer months to sleep during the colder months 500 m to 2 km below the surface.

Heterotherms are animals that sometimes regulate their body temperature and sometimes do not. They can tolerate rapid, large, repeated changes in body temperature, and can also generate significant amounts of metabolic heat at other times. Switching between poikilothermic and homeothermic characteristics may occur at different stages of daily or seasonal routines. Insects are poikilothermic but the adults of flying insects have high metabolic rates and generate metabolic heat as great or greater than homeothermic animals. They even follow Allen's rule.

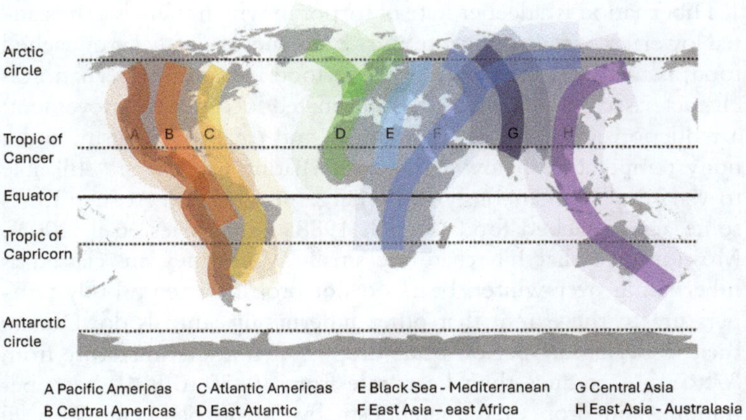

Figure 3.8 The eight major migratory bird fly ways.

Source: base map Public Domain, https://commons.wikimedia.org/w/index.php?curid=868126 – the Arctic circle, Tropic of Cancer, Equator, Tropic of Capricorn, and Antarctic circle, Antarctic land mass and currents have been added to the base map

In flight insects must maintain a constant thoracic temperature and neither lose nor retain too much heat energy. Some insects have insulation in the form of fine hairs (called pile). The ventral surface is free of pile and acts as a thermal window. Internal air sacs act as insulators, retarding heat flow from the thorax to the abdomen and as a counter-current mechanism to prevent cool blood entering the thorax. The result is that the abdomen is kept cool and the thorax warm. In warm conditions bumblebees (*Bombus* spp.) can use evaporative cooling, by regurgitating fluids, wetting the proboscis and moving it about in the air to keep cool.

Torpor is an adaptation used by a range of animals including hummingbirds (Trochidae), and small mammals such as bats (Chiroptera). Daily torpor seems to have evolved as an adaption to low temperature. For example, hummingbirds go into a state of torpor when temperatures fall overnight. Their metabolism is greatly reduced so that they do not expend more energy than is necessary. They then feed when it is warmer, and they expend less energy in so doing.

Hibernation is a deeper state of torpor in which there is a substantial lowering of metabolic rate and stored energy (body fat or cached food) is used up during a time when food is scarce. Hibernation is characterised by a cessation of coordinated locomotory movement; a reduction in heart rate, respiration, and total metabolism; and a body temperature below 10°C. Hibernating animals are still able to warm up spontaneously and emerge periodically. At such times some access cached food (French 1988; Humphries et al. 2003). Most animals that hibernate are small. While they are classed as hibernating over winter, bears do not drop their core body temperature to the extent that other hibernating animals do. During their 'hibernation' period bears drop their body temperature from 37 to 35°C, reduce their heart rate from 40–50 to 8–10 beats per minute, and drop their metabolism by 50–60 per cent. This all reduces their energy requirements, but they still lose around 20–25 per cent of their body weight during this period (Barbosa et al. 1997; Nelson 1980).

A state of dormancy similar to hibernation is used by some animals to avoid damage from high temperatures and the risk of desiccation. This period of summer dormancy is called aestivation (from the Latin *aestas* for summer) and is observed in terrestrial and aquatic invertebrates and vertebrates.

3.3.2 ADAPTATIONS TO LIGHT (INCLUDING SEASONALITY)

Some plants are adapted to live in shady conditions (termed shade tolerant plants) while others require direct sunlight. Forest floor plants in deciduous temperate forests must cope with full sunlight in the early spring and deep shade in the summer when the leaves on trees have fully opened. One adaption to these conditions is to complete their life cycle early in the growing season and remain dormant for the remainder of the year. Hence, the floor of a deciduous temperate forest may be carpeted by flowers in the early spring and then relatively bare for the rest of the year.

The daily duration of the light is termed the photoperiod. Long-day plants (perhaps more accurately short-night plants as it is the duration of darkness rather than of light that is the controlling factor) typically flower during late spring or early summer when the night length falls below the critical photoperiod for these plants.

Short-day plants (also called long-night plants) flower when the period of darkness exceeds their critical photoperiod. There are also plants that are either long-short-day plants (LSDP) or short-long-day plants (SLDP). LSDPs flower after a series of long days followed by short days whereas SLDPs flower after a series of short days followed by long days (Taiz et al. 2015). Some plants are obligate photoperiodic plants and require a long or short night before flowering, whereas others are facultative photoperiodic plants and are more likely to flower under one condition.

In animals, photoperiodicity is often referred to as seasonality. Seasonal adaptations include changes in colour of fur and feathers, changes in length and thickness of fur, migration, entry into hibernation, and courtship and breeding behaviours with associated resizing of organs. Behavioural response can be triggered by either increasing or decreasing day-length. For example, as the period of light lengthens in spring this triggers birds to breed while it is shortening day-length in autumn that triggers deer to rut.

When stimulated, photoreceptors in insect brains trigger physiological responses such as diapause induction and termination, and changing seasonal morphs (i.e. changes from eggs to larvae, to pupae, and to adults or from eggs to nymphs (which may pass through several instars) to adults depending on the life cycle of the insect) (Saunders 2012). Diapause is the term used for a delay in development in response to regular and recurring periods of adverse environmental conditions such as seasonal drought or cold weather which may also be related to reduced food supply. This behaviour allows insects, and other animals that have a seasonal dormant period, to survive these adverse conditions, and resume active growth when conditions and food supply are more favourable.

In seasonal climates, the size of the gonads in Passerine birds increases during spring. In females, ovarian follicles develop and testes grow larger in males. There are also changes in behaviour, for example males sing longer and more varied songs (Nelson 2005; Wang et al. 2020). Following reproduction, bird gonads reduce in size, thus reducing the body weight for ease of flying. Migratory behaviour in birds is also triggered by the photoperiod. With changing day length trigging appropriate seasonal movements.

Mammals, particularly those living in regions with seasons which may be hot and cold or wet and dry, exhibit seasonality which manifests in a broad spectrum of behavioural and physiological adaptations, including hibernation, seasonal migrations, coat moulting and colour change, seasonally related breeding cycles, and changes in population and community structure (Nicolas and Colyn 2003; Zimova et al. 2018).

3.3.3 ADAPTATIONS TO DROUGHT AND FLOODING

Plants adapt to moisture gradients in soil, by altering their physiology, modifying root growth and architecture, and closing or opening stomata (Gupta et al. 2020; Ahluwalia et al. 2021) (Table 3.1). Long-term responses to draught conditions include repressed shoot growth, metabolic adaptations, limited transpiration area, kernel abortion (a kernel is the seed of a grain plant such as maize (*Zea mays*) or wheat (*Triticum aestivum*)), and senescence (i.e. the condition or process of deterioration with age) (Kaur and Asthir 2017).

A few woody species can grow in very wet conditions. For example, alder (*Alnus* spp.) and willow (*Salix* spp.) thrive in wet and/or waterlogged soils. They share an adaptation to the low nitrogen levels in wet soils by forming a symbiotic relationship with nitrogen-fixing bacteria. There is also a suggestion that trees capable of fixing atmospheric nitrogen are best able to access mineral nutrients and carbon directly from bedrock (Perakis and Pett-Ridge 2019). A few woody species can grow in permanently flooded sites. For example, bald cypress (*Taxodium distichum*) and mangroves (*Rhizophora* spp.) (Figure 3.9). They do this by developing pneumatophores – specialised growths of root systems, which function to oxygenate their root system. As nutrient supply, particularly nitrogen, is also scarce in wet or water-logged soils, some plants have adapted by becoming carnivorous. They typically trap protozoans, insects, other arthropods, and, occasionally, small mammals and birds. Examples of carnivorous plants include Venus flytrap (*Dionaea muscipula*), bladderworts (*Utricularia* spp.) and pitcher plants (around 80 types found in the genus *Sarracenia, Nepenthes*, and *Darlingtonia*).

Animals possess several adaptations, which either keep water in or expel excess water from their body. This allows animals to maintain a constant internal environment in which metabolic reactions can take place. At one level, animals can move and so can search out more favourable moisture related environments. At another level,

Table 3.1 Drought tolerance mechanisms of plants, and their effects

Mechanism	*Effects*
Oxidative adjustment by producing reactive oxygen molecules including peroxides, superoxides, hydroxyl radials.	Reactive oxygen molecules are highly reactive chemicals which cause damage to cellular components including DNA, proteins, and lipids. The effects can be reduced photosynthesis, increased mutation rates, and cell death.
Decreased transpiration and stomatal conductance by stomatal closure.	Prevention of water loss due to transpiration from stomata of stems and leaves. This can cause reduced carbon dioxide and nutrient uptake, leaf shedding, and hardening of leaves (sclerophylly).
Modified root and shoot system which increase water uptake through roots and decreased transpiration rate through shoots.	Thinning of roots, short shoot growth, stunted growth, and cell damage.
Osmotic adjustment by accumulating inorganic cations (positively charged) and anions (negatively charged), organic acids, carbohydrates, amino acids, and sugars.	Detoxification of reactive oxygen molecules. Protection of cell membrane integrity. Stabilisation of enzymes or proteins.
Production of stress hormones, abscisic acid, and ethylene as a result of stimulation of plant stress signalling system.	Abscisic acid production leads to: • Increased root to shoot ratio • Altered root architecture to increase water influx into roots and reduce water loss through transpiration • Increased leaf senescence • Stomatal closing to prevent water loss • Increased adaptivity of plants. • Ethylene production leads to: • Reduced transpiration • Stimulation of fruit ripening, regulation of nodulation • Altered leaf senescence behaviour.

Source: adapted from Ahluwalia et al. (2021)

animals possess a protective outer layer that, to a greater or lesser extent, protects against passive water loss and gain. Animals also have adaptations to control the amount of moisture in the body. Single-celled animals possess contractile vacuoles, which can store and then

Figure 3.9 A mangrove swamp in Mexico illustrating plants that are adapted to living in flooded soils.

Source: photograph by Charlotte Weinberg

expel excess moisture, while higher animals have a urinary system to rid the body of excess water and solutes or to conserve them.

Drought, be it seasonal or periodic, affects animal behaviour. The African buffalo (*Syncerus caffer*) lives on the African savanna.

In the wet season, food is abundant, but during the dry seasons, as the grasses dry, food declines in terms of both quality and quantity. As the dry season progresses, the African buffalo must adapt by becoming less selective with regards to the plants it consumes (Sinclair 1997).

Dehydration may also be a problem for some animals. Camels (*Camelus* spp.) have several adaptations to minimise the amount of water they lose. They reduce unnecessary moisture loss by being able to tolerate mild hyperthermia, allowing their body temperature to rise before sweating. Their urine is very thick, syrupy and their faeces is dry. When they exhale, air passes over a mucus membrane in the nose which cools the air. This acts as a dehumidifier and water is removed from the exhaled air and reabsorbed by the body rather that it being exhaled. Finally, camels can withstand a loss of 30 per cent of their body weight in water (Osterloff 2024).

3.3.4 ADAPTATIONS TO SALINITY

Plants that are adapted to saline conditions are called halophytes (from the Ancient Greek *halas* meaning salt and *phyton* meaning plant). They can be found in aquatic habitats (including mangrove swamps, kelp forests, salt lakes, sandy and rocky shores mudflats, and salt marsh) and terrestrial environments where the soil has a salt content greater than 0.2 per cent salt (e.g. salt deserts, salt steeps, isolated areas of inland saline grassland, and places where human activities have imported salt to an area – salt is a key component in many industrial chemical processes which can result in localised soil contamination).

Saline environments, due to having a low osmotic potential, draw water out of cells, thus causing dehydration. Halophytes have adaptations – salt exclusion and salt secretion – that allow them to tolerate such conditions. Some halophytes have a higher salt concentration in their roots than in their stems and leaves. This is salt exclusion – salt is excluded from the stems and leaves by a barrier formed by the internal endodermis. Halophytes that do not have this barrier tend to secrete salts via specialised cells in their leaves. Some halophytes selectively excrete ions (e.g. excreting more sodium ions than potassium ions). Other species accumulate ions in leaf tissues that are distanced from active metabolic sites and then shed the leaves containing the accumulated ions.

Saltwater fish must be adapted to saline environments. Salmon (Salmonidae) are anadromous, meaning that they migrate up rivers to spawn and lay their eggs in the upper reaches of freshwater rivers. The eggs hatch and young salmon pass through various phases (alevins, fry, parr, and smolts) as they develop. The smolts, typically around 10 cm in length, migrate out of rivers into the ocean. The smolts develop into adults in the oceans before migrating back to the river system in which they hatched to lay and fertilise eggs thus completing the life cycle. In completing this life cycle salmon move from freshwater – where they are hypotonic – to saltwater – where they are hypertonic – and back again (Figure 3.10).

To cope with the different salinities, salmon have behavioural and physiological adaptations. For example, behaviourally, salmon do not drink water when they are in freshwater but in the dehydrating saltwater environment they drink several litres of water per day. Physiologically, kidney function differs between the two habitats. In freshwater, a salmon's kidneys produce large volumes of dilute urine, while in the ocean environment, they produce a smaller volume of concentrated urine.

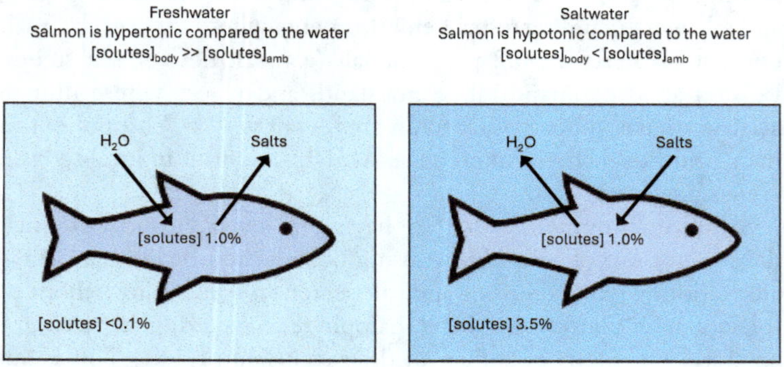

Figure 3.10 Movement of water and salts into and out of a salmon (Salmonidae) in fresh- and saltwater.

3.3.5 ADAPTATIONS TO NUTRIENT TYPE AND AVAILABILITY

3.3.5.1 Plants

Plants require nutrients, which they extract from the soil. There are nine macronutrients, so called because they are taken up in relatively large quantities: nitrogen (N), phosphorus (P), potassium (K), calcium (Ca), sulphur (S), magnesium (Mg), carbon (C), hydrogen (H), and oxygen (O). There are also eight micronutrients (or trace minerals) which are taken up in relatively small quantities: iron (Fe), boron (B), chlorine (Cl), manganese (Mn), zinc (Zn), copper (Cu), molybdenum (Mo), and nickel (Ni). Some authors (e.g. Ikram et al. 2024) also class selenium (Se) as a plant (and human) micronutrient.

One adaptation shown by plants where nutrient levels are low is to reallocate carbon to the roots moving it from the stems and leaves. This facilitates the growth of more roots to extract what nutrients are present (Bradshaw et al. 1964). A second adaptation in some plants is increased leaf longevity, so that nutrients are not lost when leaves fall. A third adaption is the formation of mutualistic relationships with microorganisms. There are two main forms of mutualistic relationships: one with nitrogen fixing cyanobacteria (aquatic environments) and bacteria (terrestrial environments), and the other with fungi (mycorrhizae). In the aquatic environment cyanobacteria are free living while in the terrestrial environment the bacteria (*Rhizobium* spp.) grow in or on the root system of certain plant species and mycorrhizal fungi are attached to roots of plants and extend out into the soil. The plant provides carbon to the cyanobacteria, bacteria, or fungi, and receive nitrogen, and in the case of the fungi other nutrients, from the symbiont.

Plants have been classified into three broad categories calcifuge (lime-hating), neutrophiles (tolerant of both acid and lime rich conditions), and calcicole (lime-loving). While the classification is named in relation to calcium, it is soil pH that is the important factor. Calcifuge plants are adapted to acidic soils, typically with a pH below 7.0. In such soils they are exposed to high concentrations of hydrogen ions (H^+); deficiencies of nitrogen, phosphorus, and magnesium; and toxicity problems associated with iron, aluminium, and other heavy metals. Calcifuge plants often have specific mechanisms to solubilise nutrients from the soil. One such mechanism is the release of organic acids from their roots. This is called organic acid exudation. These organic acids chelate (bind to) the metallic ions, making them available for absorption. Some calcifuge plants are adapted to absorb nitrogen in the form of ammonium ions (NH_4^+). Ammonium ions are more prevalent in acidic soils than nitrate ions (NO_3^-) which are more common in neutral or alkaline soils. The adaptions to toxicity issues are primarily through exclusion and internal detoxification. An example of an exclusion mechanism is the binding of metal ions in the soil to organic acids (such as citrates and malates) which are exudated by roots to form non-toxic, stable complexes. This prevents the toxic ions from being absorbed by the plant's roots, effectively creating a buffer zone in the rhizosphere (the area of soil immediately surrounding the roots). A second exclusion adaption is the development, in some calcifuge plants, of a robust root structure with a thicker endodermis (the innermost layer of the cortex). This layer acts as a barrier to the entry of toxic ions. If toxic ions do enter the plant, calcifuges can de toxify them by sequestering the ions in less sensitive parts of the plant, such as the cell walls or vacuoles, thereby, preventing them from causing damage to vital metabolic processes in the cytoplasm. These combined adaptations allow these plants to flourish where other plants would suffer from nutrient deficiencies for example heathlands and coniferous woodlands (Figure 3.11A).

Neutrophiles occupy a middle ground, preferring neutral soils where pH conditions are neither strongly acidic nor alkaline. This group includes generalist species which have broad tolerance ranges and flexible nutrient acquisition strategies. These adaptations allow neutrophiles to survival in diverse habitats. Calcicole plants are adapted to alkaline, calcium-rich soils, typically derived from limestone. They can absorb nutrients in high-pH conditions and do not have the physiological adaptations of calcifuge soils to tolerate the toxicity problems associated with acidic soils (Figure 3.11B).

3.3.5.2 Animals

Multicellular animals obtain their energy, essential nutrients, and amino acids (some of which cannot be synthesised by animals and are, therefore, termed essential amino acids) by consuming plants, fungi, and animals in various combinations. Herbivores are those

(A) (B)

Figure 3.11 Plants growing on acid (A) and alkaline (B) soils. (A) Calcifuge (lime hating) plants – common heather (light purple – *Calluna vulgaris*) and bell heather (dark purple – *Erica cinerea*) on peat soil: Muckle Roe, Shetland. (B) Calcicole (lime tolerant) plants – hawthorn (*Crataegus monogyna*), gorse (yellow flowers – *Ulex europaeus*) and various grasses.

Source: photographs by David James

animals that obtain nutrients only from plant tissue, omnivores are those that eat both plants and animals, and carnivores are those that rely on animal tissues for their nutrients.

Aquatic and terrestrial herbivores can be categorised according to the type of plant material consumed. Grazers feed on leaf materials (especially grasses); browsers on woody material, leaves, and fruits; granivores on seeds; and frugivores on fruit. There are also sap suckers and nectar consumers. Examples of sap suckers include *Sphyrapicus* spp. (a genus of sap-sucking woodpeckers found in North America) and aphids (Aphidoidea – a superfamily with a worldwide distribution). Examples of nectar consumers including hummingbirds (Trochilidae), butterflies and moths (Lepidoptera), ants (Formicidae), honeybees (Apidae), and bumblebees (*Bombus* spp.).

All herbivores face the challenge of eating plant materials which is rich in carbohydrates and low in proteins and converting it into animal tissue which is, conversely, low in carbohydrates and rich in proteins. Herbivores have developed behavioural and morphological adaptations including selective feeding on young shoots or prolonged rumination (to enhance nutrient extraction from plant matter), and specialised mouthparts (invertebrates), beaks (birds), or dentition (mammals). Herbivorous invertebrates have a range of modifications to their feeding parts, for example, for sucking, biting, and rasping plant tissues. Bird beak length and width are adapted to the type of plant material consumed. For example, birds feeding on nuts have short sturdy beaks to crack the nutshell, those feeding on insects have fine beaks to catch and hold insects, raptors (which feed on, for example, other birds and mammals) have strong hooked beaks adapted to tearing the flesh of their prey. Many mammalian herbivores have evolved bilaterally flattened, broad molar teeth with ridged surfaces for grinding leaves and stems. Mammalian herbivores eat vegetation in different ways. For example, cattle use their tongue to tear up grass, while sheep nibble the grass to a short sward. Knowledge of the different ways herbivores consume vegetation is useful in conservation practice where different animals can be employed to produce different conservation outcomes.

Omnivorous animals consume both plants and animals. Feeding habits in omnivores can vary between life stages and time of the year depending on what food is available and when. As with carnivores, omnivores have mouthparts, beaks, and detention (as relevant to the group of organisms) which are adapted to eating both plant and animal material, and a digestive system that can cope with both sources of nutrients.

There are other feeding strategies, including detritivores. Detritivores feed on fallen leaves, twigs, other dead wood, and dead organic material including animal carcasses and faeces (feeding on faeces is termed coprophagy). Most detritivores are invertebrates. In addition to adaptations in the mouthparts, they also rely on mutualistic relationships with microorganisms, which break down and enrich the dead material to the extent that the detritivore can consume the material more easily.

3.3.6 ADAPTATIONS TO NOISE

For an animal to make informed behavioural decisions it must be able to accurately receive information – primarily visual, auditory, and olfactory though some animals also possess other senses such as a sensitivity to electric fields (e.g. electric eel (*Electrophorus* spp.)). This information is used by animals to detect the presence of predators, prey, and potential mates, and about the environment (Endler 1993).

Regarding auditory signals and cues, animal behaviour is shaped by both natural noise and the, frequently louder and persistent, noise from anthropogenic sources (e.g. traffic, aircraft, and industrial equipment) (Kight and Swaddle 2011; Gomes et al. 2021). When anthropogenically created noise overlaps natural noise, the natural noise may be distorted or masked, thereby inhibiting effective information flow (Brumm and Zollinger 2013; Endler 1992).

Anthropogenic underwater sound from commercial shipping, seismic surveys, naval sonar, and offshore construction affects aquatic life at multiple levels. Moderate continuous noise masks communication and echolocation calls, disrupting predator–prey interactions and reducing foraging efficiency across diverse taxa. High-intensity impulsive sounds can induce temporary or permanent hearing loss, tissue damage, and chronic stress in fish, invertebrates, and marine mammals, compromising survival and growth (Hawkins and Popper 2017). These individual-level impacts can alter reproductive behaviour, decrease recruitment rates, and drive shifts in community composition with long-term consequences for ecosystem structure and function (see Chapter 6) (Popper et al. 2023).

Behavioural adaption to anthropogenic noise is seen across numerous taxa. Some birds delay their dawn chorus until the noise from aircraft or morning insect chorus reduces (Gil et al. 2015; Stanley et al. 2016) and brown tree frogs (*Litoria ewingii*) increase the pitch of their songs to avoid frequency overlap with traffic noise the (Higham et al. 2021). These examples all relate to signallers enhancing the production and transmission of information amid noise (sound-to-noise ratio). The funnel-weaving spider (*Agelenopsis pennsylvanica*) has an adaption that influences information reception. Pessman and Hebets (2025) report that funnel spiders in rural and urban settings build their webs differently. The webs vibrate in response to a prey item being caught, the approach of a mate, and environmental noise. Under loud environmental conditions the webs of rural funnel spiders were found to retain more energy over a longer distance across a narrow frequency range (350–600 Hz). This is, potentially to improve the reception of relevant prey and mate cues. Urban funnel spider webs, conversely, lose energy over a short distance across a broader frequency range (300–1,000 Hz). This adaption is, potentially, to prevent sensory overload from constant, high-amplitude urban noise.

3.3.7 EXTREMOPHILES

It was only in the 1980s and 1990s that biologists began to find microbial life surviving, and in some cases thriving, under conditions which until then it had been considered could not be tolerated by any organism (Figure 3.12). These microbes are examples of extremophiles (from the Latin *extremus* (extreme) and Ancient Greek *philia* (fondness for)). Extremophiles have been found in environments with extreme temperatures, pH levels, salinity, pressure, and radiation. For example, thermophiles (i.e. species capable

Figure 3.12 Example of extremophiles in hot springs and geysers in Yellowstone National Park, USA. (A) Porcelain Basin with steaming vents. (B) Giant Prismatic Spring (mean temperate 70°C). (C) Morning Glory Pool (mean temperate 70.7°C, mean pH 7.6, and mean conductivity 2098 μS/cm). (D) an acidic, hot spring – colourless and yellow thermophiles grow in the hottest water. Orange, brown, and green thermophiles grow in cooler waters.

Source: photographs by Rob Young

of tolerating high temperatures) flourish in hydrothermal vents at temperatures exceeding 100°C. *Methanopyrus kandleri* (an archaean) holds the record for any life form at 122°C (Rappaport et al. 2025). Halophiles, another example of extremophiles) are species capable of tolerating high salt concentrations such as the roughly 35 per cent salinity of the Dead Sea and other highly saline lakes.

Many extremophiles are archaea, though some bacteria and eukaryotes display similar resilience to extreme environmental conditions (Table 3.2).

Extremophiles exhibit a range of adaptions to their environments including osmoprotectants (to allow survival in highly saline environments) and efficient DNA repair mechanisms (to

Table 3.2 Examples of extremophiles

Extremophile	*Comments*
Incendiamoeba cascandensis	An amoeba whose scientific name translates as 'fire amoeba from the cascades' discovered in Lassen Volcanic National Park, California, USA. It was found living at 57°C and found to be able to live at 64° At 70°C, the amoeba formed a dormant 'cyst' that was reactivated when temperatures cooled (Rappaport et al. 2025).
Giant tube worm (*Riftia pachyptila*)	Found living in conditions of high pressure, high heat, and no sunlight near deep-sea hydrothermal vents. They have symbiotic extremophilic bacteria in their midgut. The bacteria, which may account for up to half of the worm's weight, use chemosynthesis (see Chapter 6) to turn oxygen, hydrogen sulphide, and carbon dioxide into organic molecules that the worm can use.
Grylloblattidae	A family of psychrophilic (an organism that grows best at temperatures close to freezing) insects that are found in cold environments (1–4°C) (e.g. mountain tops, glaciers, and ice sheets). When temperatures drop below freezing, and to avoid ice-crystals forming in their bodies, these insects burrow down through the snow and stay near the soil.
Loricifera	Microscopic marine sediment-dwelling animals. The environment is characteristically cold, high pressure, low light, low oxygen, sulphur rich, salt-saturated brine. Loricifera have hydrogenosomes that, unlike mitochondria, do not require oxygen to produce energy.

(*Continued*)

Table 3.2 (Continued)

Extremophile	*Comments*
Snoticle or Snottite	Colonies of microbes comprising the bacterium *Acidithiobacillus thiooxidans* (> 70 per cent of cells), an archaeon (microorganisms that resemble bacteria but differ from them in their genetic makeup and cell structure) (>15 per cent), and a bacterium in the Acidimicrobiaceae family (>5 per cent) (Jones et al. 2012). These colonies survive environments that are highly acidic and, for other organisms, highly toxic. They use chemosynthesis to turn volcanic sulphur compounds into energy and sulphuric acid waste.
Tardigrades	Tardigrades are eight-legged, segmented invertebrates. Technically, they are extremotolerant rather than extremophilic. Tardigrades are known to survive temperatures from –270°C to +180°C, pressures six times greater than that found in the deepest parts of the ocean, very high doses of radiation, and the vacuum of space. When they cannot tolerate an environment, they expel most of their body water and enter a 'tun state' – a state in which the body is extremely dehydrated and crinkled – this process is termed cryptobiosis. In the tun state, they produce proteins and sugars, which stabilise cellular structures and prevent molecular damage from high levels of ionising radiation. They also have robust DNA repair systems that reverse any damage incurred during these extreme episodes.

allow survival in high radiation environments). A key to adaptation which allows thermophiles to live in high temperature conditions is the structure of their proteins. The amino acid composition of proteins in thermophiles is different from that in non-thermophiles. This difference is expressed in the way proteins fold under particular conditions and how denaturation of proteins is prevented thereby rendering enzymes (which are proteins) heat stable (Reed et al. 2013).

An extension of the study of extremophiles is the search for life on other planets where conditions are as, or more, extreme than those found on Earth. The study of extremophiles also has applications in biotechnology, and environmental remediation.

3.4 CASE STUDY – URBANISATION

3.4.1 GROWING URBAN AREAS

The urban landscape is becoming the most familiar environment for most of humanity (Niemelä et al. 2011) (Figure 3.13). The global urban population is projected to increase to peak at 10.3 billion in 2084 (United Nations 2024). The mean annual growth in population is 1.58 per cent and the average annual urban land expansion rate in a city with more than a million inhabitants is 1.08 per cent (Mahtta et al. 2022). Gao and O'Neil (2020) built projections, based on 15 diverse datasets, indicating that the amount of urban land on Earth by 2100 could range from about 1.1 million to 3.6 million km^2 across five different socioeconomic scenarios. This is roughly 1.8–5.9 times the global total urban area in 2000 (which was about 0.6 million km^2) (Gao and O'Neil 2020). Global per capita urban land is forecast to more than double from $100\,m^2$ in 2000 to $246\,m^2$ in 2100. The fastest urban land expansion is projected to occur in Africa and Asia, though the developed world experiences a similarly large amount of new development (Gao and O'Neil 2020). Accommodating this growing urban population

Figure 3.13 The central business district of Perth, Western Australia, taken from the city's Kings Park (see Department of Biodiversity, Conservation and Attractions 2026).

Source: photograph by Mike Allen

requires the enlargement of existing urban centres and the construction of new centres. These projections illustrate the growth of the urban environment and associated with this growth is a greater impact.

Urbanisation is a complex socio-economic process with numerous consequences for biodiversity, the environment, and society (United Nations 2018). Drivers of change within an urban setting include pollution, land cover change, biotic invasion, climate, and human activities (Ruas et al. 2022).

Urbanisation affects biogeochemical cycles, regional to global climate, hydrological systems, and biodiversity. It is strongly linked to higher per capita urban greenhouse gas emissions, degradation and fragmentation of natural habitats, biodiversity loss; inefficient use of natural resources; loss of agricultural lands; the heat-island effect; pollution (of water and air and by noise and light); and the introduction of non-native species (Grimm et al. 2008; Parris 2016).

The climate in an urbanised area differs from that in the surrounding rural landscape. The key drivers of the differences are the nature of the material used in the construction of roads, buildings, and other infrastructure. These materials tend to absorb more solar energy than trees, grass, and bare earth. The effect is that heat energy is stored in these materials during sunny periods and released back into the atmosphere during the night. Within urban areas there is also a significant use of machinery, vehicles, heating and cooling (as appropriate to the climate) inside buildings, and of fossil fuels for preparing and storing food all of which release heat, greenhouse gases, and pollutants into the atmosphere. The effect of all these energy changes gives rise to an urban heat island (UHI) (i.e. the temperature in an urban area is higher than in surrounding rural areas) (Table 3.3).

Urbanisation leads to the creation of novel ecosystems that exhibit novel abiotic and biotic characteristics and pose a challenge to many organisms (McKinney 2008). Only species that can cope with or adapt to this novel set of challenging urban conditions can survive in cities. While generalisations can be drawn, responses to urbanisation of local biodiversity are also a function of the biome and climate relevant to each city.

Table 3.3 Generalised changes in climatic characteristics due to urbanisation

Increasing due to urbanisation		*Decreasing due to urbanisation*	
Variable	*Higher/more (%, unless stated otherwise)*	*Variable*	*Lower/less (%, unless stated otherwise)*
Temperature		*Relative humidity*	
Annual mean	0.5–3.0°C	Annual mean	6
Summer maximum	1.0–3.0°C	Summer	2
Winter minimum	2.5–4.0°C	Winter	8
Cloudiness		*Solar radiation*	
Clouds	5–10	Total, horizontal	12
Summer Fog	100	Winter, ultraviolet	30
Winter Fog	30	Summer, ultraviolet	5
Air turbulence		*Sunshine*	
Calms	5–20	Duration	5–15
Precipitation		*Wind speed*	
Amounts	5–10	Annual mean	20–30
Days with < 5 mm	10	Extreme gusts	10–20
Thunderstorms	10–5		
Snowfall		*Snowfall*	
Downwind of city	10	Inner city	5–10

Source: based on Bridgman et al. (1995)

3.4.2 PLANTS IN URBAN AREAS

The responses of plants to urbanisation can be complex and varied (Figure 3.14). An individual plant response may be the result of one urban driver (e.g. land cover change, biotic invasion, climate, human activities, and pollution) or of several drivers working together. In general, urbanisation mainly affects plants negatively both in megacities (i.e. a city with a population over ten million people) and smaller cities.

Some plants thrive as a result of urbanisation. For example, in countries where salt is spread on roads to prevent icing, roadside

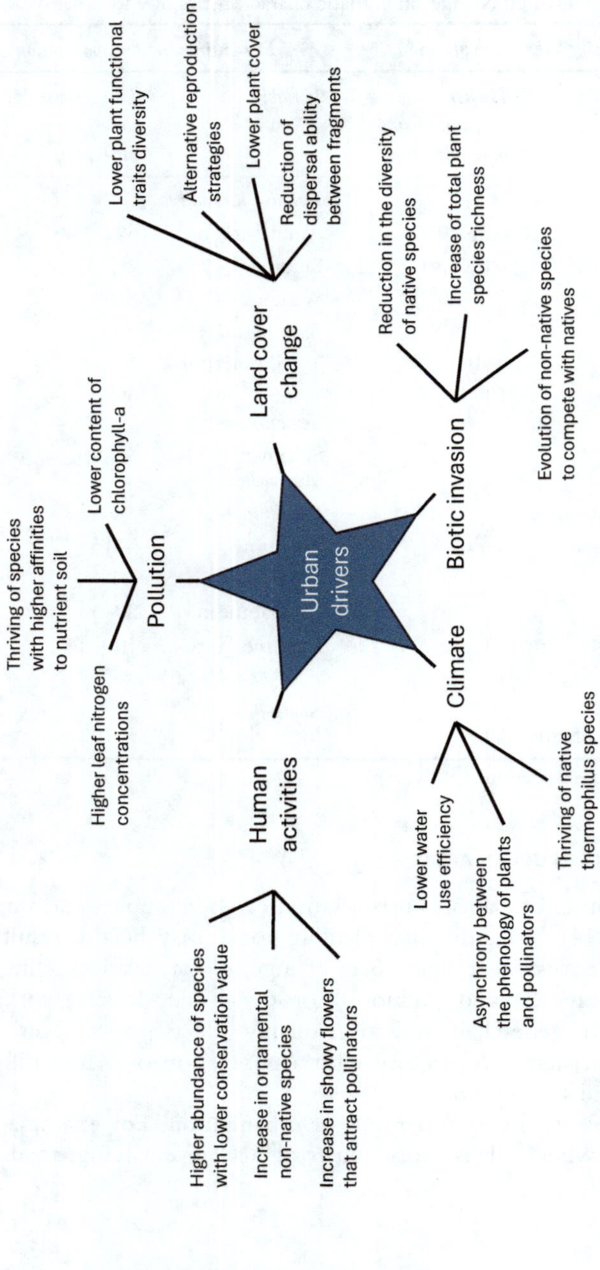

Figure 3.14 Adaptations shown by plants to urbanisation.

Source: based on Ruas et al. (2022)

verges and motorway central reservations become habitats with a high salt concentration. Danish scurvy grass (*Cochlearia danica*) outcompetes less salt-tolerant species in these habitats and forms dense fringes (which turn white as the plant flowers in spring) along major routes. Wind turbulence from fast-moving vehicles further aids seed dispersal.

3.4.3 ANIMALS IN URBAN AREAS

Research indicates that urban environments drive adaptive responses in behaviour including changes in home range and diet preference, shifts in activity budget and vigilance, decreased flight initiation distance, and increased nocturnal activity. Some urban mammal species even demonstrated the ability to modulate behaviours based on environmental cues (Ritzel and Gallo 2020). Animal behavioural responses can be categorised in terms of timescale and permanence, as either regulatory, acclimatory, or developmental (Lopez-Sepulcre and Kokko 2012; McDonnell and Hahs 2015). These adaptive responses may complement species survivability in some cases while being detrimental in others (Ritzel and Gallo 2020) (Table 3.4).

The overall effects of urbanisation on biodiversity are negative (McKinney 2008; Fenoglio et al. 2020; Theodorou et al. 2020). The influence of urban environments on birds was reviewed by Isaksson (2018) who found evidence of behavioural and physiological changes related to the urban environment (Table 3.5), and on

Table 3.4 Animal behavioural responses to the urban environment

Behavioural responses	*Examples*	*Time scale and permanence*
Regulatory	Changes in alert behaviour including harm avoidance and decreased flight initiation distance (FID)	Develop quickly in seconds to hours
Acclimation	Adjustments in social structure and territoriality	Develops gradually over days and weeks
Developmental	Physiological changes and behavioural syndromes including neophilia and boldness	Permanent developmental response

Table 3.5 Avian behavioural and physiological changes related to the urban environment

Urban driver	*Behavioural and physiological responses*
Artificial light at night (ALAN)	Earlier dawn singing earlier Changed mating behaviours Reduced melatonin which might affect sleep Changed movement routes due to navigation and orientation system of nocturnally migrating birds being influenced by light (e.g. 'trapping' of birds, i.e. birds are attracted to beams of light and, while in the beam, they get disorientated and fly around in circles within the beam) Earlier reproduction in seasonal breeders
Chemical pollution	Changed physiological
Noise	Reduced flight-initiation distance as the sound of a threat is not detected above the background noise until the threat is nearer to the animal Shifts in the frequency and timing of vocal communication (a sift to higher frequency songs which can be heard above the generally lower pitch of the background urban noise, and vocalising when the urban noise is quieter)
Novel foods	Reduced growth and development, immune system, and yellow-red plumage pigmentation (which can affect mate selection as colours that are less bright are less attractive to potential mates) due to the typically lower level of carotenoids in foods in urban areas Changed beak morphology according to food type Greater fat storage Heightened inflammatory responses (linked to fatty acid composition of foods) Changed metabolic rate

Source: based on Isaksson (2018)

mammals by Ritzel and Gallo (2020) (Table 3.6). The direction of the responses in Table 3.6 (e.g. an increase or decrease) is dependent on the environmental factors in each city. For example, where there is greater food available in an urban area compared with an adjacent rural area, the foraging time for a mammal will be less in the urban aera and its home range will be smaller than in the rural area.

Duarte and Young (2011) reported on urban marmosets (*Callithrix penicillata*) in Belo Horizonte City, Minas Gerais, Brazil, and specifically their behavioural response to avoid predation by domestic cats (Box 3.1).

Table 3.6 Behaviour changes found in urban mammals grouped by animal order. The direction of the response (i.e. increase or decrease) depends on the conditions in a specific urban area as compared with that in a nearby rural area. Abbreviations: AD = alert distance (the distance at which an animal begins to exhibit alert behaviours towards an approaching human or predator); GUD = giving-up density (a metric used to measure foraging behaviour in animals which represents to food density within a patch at which an animal decides to leave that area and move to another); FID = flight initiation distance (the distance at which an animal begins to flee from an approaching threat); VED = vertical escape distance (the distance an animal can move vertically to escape a potential predator)

Order	*Behaviours affected by urbanisation*
Artiodactyla (even-toed ungulates)	Anti-predator avoidance, FID (flight initiation distance), home range, nocturnality, resource selection, spatial patterns, travel distance vocalisation
Carnivora (carnivores)	Activity budget, AD (alert distance), anti-predator, avoidance, boldness, den selection, denning time, diet preference, dispersal, escape activity, exploratory, fecundity, FID (flight initiation distance), foraging/hunting, home range, movement speed, nocturnality, resource selection, spatial patterns, territoriality, time encamped, time foraging, time/distance travelling, vigilance/caution
Chiroptera (bats)	Foraging time, home range
Didelphimorphia (opossums)	Home range, travel distance
Diprotodontia (marsupials with two forward teeth – restricted to Australasia includes kangaroos, wallabies, possums, koala, and wombats)	Diet preference, home range, vigilance/caution
Eulipotyphla (insectivores including hedgehogs, moles, shrews, and solenodons)	Avoidance, diel activity (i.e. active over the whole 24-hour period of a day), spatial patterns

(*Continued*)

Table 3.6 (Continued)

Order	*Behaviours affected by urbanisation*
Lagomorpha (rabbits and hares, and pikas)	Activity budget, anti-predator, latrine use, spatial patterns, time resting, vigilance/caution
Peramelemorphia (includes the bandicoots and bilbies – all endemic to Australia-New Guinea)	Home range
Primate (primates)	Activity budget, anti-predator, conspecific tolerance, diet preference, exploratory, foraging/hunting, grooming, play, problem-solving, time feeding, time resting, time travelling, vigilance/caution
Rodentia (rodents)	Activity budget, AD (alert distance), aggression, alarm behaviour, alert response, anti-predator, concealment distance, den selection, denning time, diet preference, dispersal, diurnality, FID (flight initiation distance), foraging time, foraging/hunting, GUD (giving-up density), home range, latency, resource selection, spatial patterns, VED (vertical escape distance), vigilance/caution

BOX 3.1 BEHAVIOURAL ADAPTATIONS OF MARMOSETS (*CALLITHRIX PENICILLATA*) IN BELO HORIZONTE CITY, MINAS GERAIS, BRAZIL (BASED ON DUARTE AND YOUNG 2011)

Primates select sleeping sites based on factors including:

- safety from predators,
- comfort and hygiene,
- social opportunities,
- thermoregulation,
- absence of parasites,
- territoriality and resource availability,
- proximity to water and food, or
- shelter from rain.

Species that suffer high predator pressure (e.g. marmosets), especially those with nocturnal predators (e.g. cats) are expected to be highly selective in sleeping site selection. Marmosets are particularly susceptible to cats that can access their sleeping sites, as they do not wake easily and are often sluggish soon after waking (Erkert 1989).

Marmosets may reduce their predation risk at sleeping sites by:

- selecting sleeping sites with particular physical characteristics (see below)
- modifying their retirement behaviour, (e.g. quickly entering a sleeping site or not scent marking nearby); or
- through their temporal and spatial patterning of sleeping site reuse

Duarte and Young (2011) studied the sleeping sites of marmosets in an 18-ha area in the Municipal Park of Belo Horizonte City, Minas Gerais, Brazil where there was one group of marmosets and at least 115 domestic cats living. The park contains 3,777 trees of 275 species. Between January and April 2009, Duarte and Young observed predation of three juvenile marmosets by domestic cats and a mean of 3.24 attempted predation events per day. They located and recoded the characteristics of 12 marmoset sleeping sites in the park.

Duarte and Young found that of the 275 species marmosets chose only six different species – three palms (Arecaceae), two eucalyptuses (*Eucalyptus*) and one fig (*Ficus*) – in which to sleep and a total of 12 different sleeping trees. Cats were also observed sleeping in trees in the park. Duarte and Young found that the marmoset sleeping sites shared certain characteristics compared with those in which cats slept: trees in which marmosets slept were tall (22.49 m for marmosets compared with 12.22 m for cats), had high first branches (as measured from ground level – 10.36 m for marmosets compared with 2.46 m for cats), had smooth or thorny bark; and could be reached only by jumping from adjacent trees.

This suggests that marmosets have adapted their sleeping behaviours, particularly their sleeping site selection, to avoid this predator, one that they would not encounter in high numbers outside a city.

Knowing how plants and animals respond to new environments is important in understanding the impacts of environmental change, be this climate, pollution, land use, or the arrival of new species. While some species can adapt to the changes they meet, others cannot. When an individual is not able to adapt, there are only two outcomes: the individual moves to where it can survive, or it dies. Non-motile species lack the first option. Even motile species may not be able to move far and/or fast enough to find suitable habitat.

Species can be classified as urban avoiders, urban tolerators, urban adaptors, and urban exploiters. Understanding these categories helps ecologists assess the impact of urbanisation on biodiversity and develop strategies for conservation and urban planning. Urban avoiders are species that struggle to survive in cities. These organisms typically require large, undisturbed natural areas and are sensitive to urbanisation. Examples include certain forest-dwelling birds (e.g. spotted owl (*Strix occidentalis*) and large mammals (e.g. wolf (*Canis lupus*) and moose (*Alces alces*). Urban tolerators are species that can persist in urban areas but do not necessarily thrive. They may utilise green spaces, parks, and suburban areas while avoiding the most densely developed regions. These species exhibit some behavioural and physiological flexibility, allowing them to coexist with human activity without fully integrating into urban ecosystems (e.g. many songbirds (Passeri) and small mammals (e.g. hedgehog (Erinaceidae)). Urban adaptors are species that successfully adjust to urban environments by modifying their behaviours and resource use. They often exploit human-made structures for shelter and food, demonstrating a high degree of flexibility. Such species can be found in both natural and urban settings, making use of gardens, rooftops, and artificial water sources. Examples include feral pigeon (*Columba livia*), raccoon (*Procyon lotor*), and red fox (*Vulpes vulpes*), which have learned to navigate and benefit from urban landscapes. Urban exploiters are species that thrive in cities, often becoming more abundant than in their natural habitats. They rely heavily on human resources, such as food waste and artificial nesting sites, and may even depend on urban environments for survival. These species often exhibit bold behaviours and high reproductive rates, allowing them to dominate urban ecosystems. Examples include cockroaches (Blattodea), house sparrow (*Passer domesticus*), rats (Rodentia), and South American coati (*Nasua nasua*) (Figure 3.15).

(A) (B)

Figure 3.15 Urban adapters. (A) House sparrow (*Passer domesticus*). (B) Female South America coati (*Nasua nasua*) raiding a bin in Mangabeiras Park, Belo Horizonte, Brazil.

Sources: (A) photograph by Luke Blazejewski; (B) photograph by Rob Young

3.5 SELF-STUDY SUGGESTIONS

3.5.1 SELF-STUDY QUESTIONS

1. How is the concept of a limiting factor helpful in understanding the distribution of species across the globe?
2. Why is too much carbon dioxide a problem for plants?
3. How do plants and animals show stress if they are in an environment where the parameters are outside the optimal zone?
4. What are the four main ways in which a plant can adapt to its environment?
5. How does drought affect plants?
6. How are plants adapted to different climates?
7. What are the differences and similarities between poikilotherms, homeotherms, and heterotherms?
8. How do reptiles regulate their body temperature?
9. How does the counter-current theory explain why birds can stand in 4°C water without losing core body temperature?
10. How does light influence the behaviour of animals?
11. What are the effects of artificial light at night (ALAN) on insects, birds, and mammals?
12. How are animals adapted to eating plants and to eating other animals?
13. How may organisms respond to the physiological challenges of the urban environment?

3.5.2 A DEEPER DIVE

A. In the context of where you live or are studying ecology, how are plants and animals adapted to that local environment?
B. List the carnivores, omnivores, and herbivores that you encounter in your local area.
C. In your local environment, what evidence do you see of niche partitioning?

3.6 REFERENCES

Ahluwalia, O., Singh, P.C. and Bhatia, R. (2021). 'A review on drought stress in plants: Implications, mitigation and the role of plant growth promoting rhizobacteria'. *Resources, Environment and Sustainability* 5: 100032. https://doi.org/10.1016/j.resenv.2021.100032.

Allen, J.A. (1877). 'The influence of physical conditions in the genesis of species'. *The Radical Review* 1: 108–140.

Barbosa, P, Farley, S. and Robbins, C. (1997). 'Whole body urea cycling and protein turnover during hyperphagia and dormancy in growing bears (*Ursus americanus* and *U. actos*)'. *Canadian Journal of Zoology* 75: 2129–2136.

Blackman. F.F. (1905). 'Optima and limiting factors'. *Annals of Botany* 19(2): 281–296. https://doi.org/10.1093/oxfordjournals.aob.a089000.

Bradshaw, A.D., Lodge, R.W., Jowett, D. and Snaydon, R.W. (1964). 'Experimental investigations into the mineral nutrition of several grass species. IV. Nitrogen level'. *Journal of Ecology* 52: 655–676.

Bridgman, H., Warner, R. and Dodson, J. (1995). *Urban Biophysical Environments*. South Melbourne, Australia: Oxford University Press.

Brumm, H. and Zollinger, S.A. (2013). 'Avian vocal production in noise' In H. Brumm (ed.), *Animal Communication and Noise*, 187–227. New York: Springer. https://doi.org/10.1007/978-3-642-41494-7.

Churchfield, S., Rychlik, L. and Taylor, J.R.E. (2012). 'Food resources and foraging habits of the common shrew, *Sorex araneus*: does winter food shortage explain Dehnel's phenomenon?'. *Oikos* 121(10): 1593–1602. https://doi.org/10.1111/j.1600-0706.2011.20462.x.

Department of Biodiversity, Conservation and Attractions. (2026). 'Kings Park and Botanic Garden'. Retrieved from www.dbca.wa.gov.au/botanic--gardens-and-parks-authority/kings-park-and-botanic-garden (accessed 25 May 2026).

Duarte, M.H.L. and Young, R.J. (2011). 'Sleeping site selection by urban marmosets (*Callithrix penicillata*) under conditions of exceptionally high predator density'. *International Journal of Primatology* 32: 329–334. https://doi.org/10.1007/s10764-010-9468-5.

Endler, J.A. (1992). 'Signals, signal conditions, and the direction of evolution'. *The American Naturalist* 139: S125–S153. https://doi.org/10.1086/285308.

Endler, J.A. (1993). 'Some general comments on the evolution and design of animal communication systems'. *Philosophical Transactions of the Royal Society B* 340: 215–225. https://doi.org/10.1098/rstb.1993.0060.

Erkert, H. G. (1989). 'Characteristics of the circadian activity rhythm in common marmosets (*Callithrix j. jacchus*)'. *American Journal of Primatology* 17, 271–286.

Fenoglio, M.S., Rossetti, M.R. and Videla, M. (2020). 'Negative effects of urbanization on terrestrial arthropod communities: a meta-analysis'. *Global Ecologyl and Biogeography* 29: 1412–1429. https://doi.org/10.1111/geb.13107.

French, A.R. (1988). 'The patterns of mammalian hibernation'. *Scientific America* 76: 569–575.

Gao, J. and O'Neill, B.C. (2020). 'Mapping global urban land for the 21st century with data-driven simulations and shared socioeconomic pathways'. *Nature Communications* 11: 2302. https://doi.org/10.1038/s41467-020-15788-7.

Gil, D., Honarmand, M., Pascual, J., Pérez-Mena, E. and Macías Garcia, C. (2015). 'Birds living near airports advance their dawn chorus and reduce overlap with aircraft noise'. *Behavioral Ecology* 26: 435–443. https://doi.org/10.1093/beheco/aru207.

Gomes, D.G.E., Francis, C.D. and Barber, J.R. (2021). 'Using the past to understand the present: coping with natural and anthropogenic noise'. *BioScience* 71: 223–234. https://doi.org/10.1093/biosci/biaa161.

Grimm, N.B., Faeth, S.H., Golubiewski, N.E., Redman, C.L., Wu, J., Bai, X. and Briggs, J.M. (2008). 'Global change and the ecology of cities'. *Science* 319: 756–760. https://doi.org/10.1126/science.1150195.

Gupta, A, Rico-Medina, A. and Caño-Delgado, A.I. (2020). 'The physiology of plant responses to drought'. *Science* 368 (6488): 266–269. https://doi.org/10.1126/science.aaz7614.

Hawkins, A.D. and Popper, A.N. (2017). 'A sound approach to assessing the impact of underwater noise on marine fishes and invertebrates'. *ICES Journal of Marine Science* 74(3): 635–651. https://doi.org/10.1093/icesjms/fsw205.

Higham, V., et al. (2021). 'Traffic noise drives an immediate increase in call pitch in an urban frog'. *Journal of Zoology* 313: 307–315. https://doi.org/10.1111/jzo.12866.

Hill, R.W. (1976). *Comparative Physiology of Animals: An Environmental Approach*. New York: Harper and Row.

Humphries, M.M., Thomas, D.W. and Kramer. D.L. (2003). 'The role of energy availability in mammalian hibernation: a cost-benefit approach'. *Physiological and Biochemical Zoology: Ecological and Evolutionary Approaches* 76(2): 165–179. https://doi.org/10.1086/367950.

Ikram, S., Li, Y., Lin, C., Yi, D., Heng, W., Li, Q., Tao, L., Hongjun, Y. and Weijie, J. (2024). 'Selenium in plants: a nexus of growth, antioxidants, and phytohormones'. *Journal of Plant Physiology* 296: 154237. https://doi.org/10.1016/j.jplph.2024.154237.

Isaksson, C. (2018). 'Impact of urbanization on birds'. In: D. Tietze. (ed) *Bird Species: Fascinating Life Sciences*. Springer, Cham. https://doi.org/10.1007/978-3-319-91689-7_13.

Jones, D.S., Albrecht, H.L., Dawson, K.S., Schaperdoth, I., Freeman, K.H., Pi, Y., Pearson, A. and Macalady, J.L. (2012). 'Community genomic analysis of the extremely acidophilic sulfur-oxidizing biofilm'. *ISME Journal* 6 (1): 158–170. https://doi.org/10.1038/ismej.2011.75.

Kaur, G. and Asthir, B. (2017). 'Molecular responses to drought stress in plants (review)'. *Biologia Plantarum* 61(2): 201–209. https://doi.org/10.1007/s10535-016-0700-9.

Kight, C.R. and Swaddle, J.P. (2011). 'How and why environmental noise impacts animals: an integrative, mechanistic review'. *Ecology Letters* 14: 1052–1061. https://doi.org/10.1111/j.1461-0248.2011.01664.x.

Liebig, J. (1840) *Organic Chemistry, in its Applications to Agriculture and Physiology*. London: Taylor and Walton.

Lopez-Sepulcre, A. and Kokko, H. (2012). 'Understanding behavioural responses and their consequences'. In U. Candolin and B.B.M. Wong (eds), *Behavioural Responses to a Changing World: Mechanisms and Consequences, 280.* Oxford: Oxford University Press.

Mahtta, R., Fragkias, M., Güneralp, B., Mahendra, A., Reba, M., Wentz, E.A. and Seto, K.C. (2022). 'Urban land expansion: the role of population and economic growth for 300+ cities'. *Urban Sustainability* 2: 5 https://doi.org/10.1038/s42949-022-00048-y.

McDonnell, M.J. and Hahs, A.K. (2015). 'Adaptation and adaptedness of organisms to urban environments'. *Annual Review of Ecology, Evolution, and Systematics* 46: 261–280. https://doi.org/10.1146/annurev-ecolsys-112414-054258.

McKinney, M.L. (2008). 'Effects of urbanization on species richness: a review of plants and animals'. *Urban Ecosystems* 11: 161–176. https://doi.org/10.1007/s11252-007-0045-4.

Nelson, R.A. (1980). 'Protein and fat metabolism in hibernating bears'. *Federation Proceedings* 39: 2955–2958.

Nelson, R.J. (2005). *An Introduction to Behavioral Endocrinology*. Sunderland, MA: Sinauer Associates.

Nicolas, V. and Colyn, M. (2003). 'Seasonal variations in population and community structure of small rodents in a tropical forest of Gabon'. *Canadian Journal of Zoology.* 81: 1034–1046. https://doi.org/10.1139/z03-092.

Niemelä, J., et al. (2011). *Urban Ecology: Patterns, Processes, and Applications*. Oxford: Oxford University Press. https://doi.org/10.1093/acprof:oso/9780199563562.001.0001.

Osterloff, E. (2024). 'How do camels survive in deserts?'. Retrieved from www.nhm.ac.uk/discover/how-do-camels-survive-in-deserts.html (accessed 23 August 2024).

Parris, K.M. (2016). *Ecology of Urban Environments*. Oxford: Wiley Blackwell.

Perakis, S. and Pett-Ridge, C. (2019). 'Nitrogen-fixing red alder trees tap rock-derived nutrients'. *PNAS* 116(11): 5009–5014. https://doi.org/10.1073/pnas.1814782116.

Pessman, B.J. and Hebets, E.A. (2025). 'Web transmission properties vary with a spider's past and current noise exposure'. *Current Biology* 35: 1706–1715. https://doi.org/10.1016/j.cub.2025.02.041.

Popper, A.N., Sisneros, J., Hawkins, A.D. and Thomsen, F. (eds). (2023). *The Effects of Noise on Aquatic Life: Principles and Practical Considerations*. Cham: Springer. https://doi.org/10.1007/978-3-031-10417-6.

Rappaport, H.B., et al. (2025) 'A geothermal amoeba sets a new upper temperautre limit for eukaryotes'. *BioRxiv preprint*. https://doi.org/10.1101/2025.11.24.690213.

Reed, C.J., Lewis, H., Trejo, E., Winston, V. and Evilia. C. (2013). 'Protein adaptations in archaeal extremophiles'. *Archaea* 2013: 373275. https://doi.org/10.1155/2013/373275.

Ritzel, K. and Gallo, T. (2020). 'Behavior change in urban mammals: a systematic review'. *Frontiers in Ecology and Evolution* 8: 576665. https://doi.org/10.3389/fevo.2020.576665.

Ruas, R.D.B., Costa, L.M.S. and Bered, F. (2022). 'Urbanization driving changes in plant species and communities – a global view'. *Global Ecology and Conservation* 38: e02243. https://doi.org/10.1016/j.gecco.2022.e02243.

Ryding, S., Klaassen, M., Tattersall, G.J., Gardner, J.L. and Symonds, M.R.E. (2021). 'Shape-shifting: changing animal morphologies as a response to climatic warming'. *Trends in Ecology & Evolution* 36(11): 1036–1048. https://doi.org/10.1016/j.tree.2021.07.006.

Saunders, D.S. (2012). 'Insect photoperiodism: seeing the light'. *Physiological Entomology* 37(3): 207–218. https://doi.org/10.1111/j.1365-3032.2012.00837.x.

Schmidt-Nielsen, K. (1964). *Animal Physiology: Adaptation and Environment* (5th edition). New York: Cambridge University Press.

Shelford, V.E. (1913). *Animal Communities in Temperate America: as illustrated in the Chicago Region*. Chicago, IL: The University of Chicago Press.

Sinclair, A.R.E. (1997). *The African Buffalo: A Study of Resource Limitation of Populations*. Chicago, IL: University of Chicago Press.

Stanley, C.Q., Walter, M.H., Venkatraman, M.X. and Wilkinson, G.S. (2016). 'Insect noise avoidance in the dawn chorus of Neotropical birds'. *Animal Behaviour* 112: 255–265. https://doi.org/10.1016/j.anbehav.2015.12.003.

Taiz, L., Zeiger, E., Møller, I. and Murphy, A. (2015). *Plant Physiology and Development* (6th edition). Sunderland, MA: Sinauer Associates.

Theodorou, P., et al. (2020). 'Urban areas as hotspots for bees and pollination but not a panacea for all insects'. *Nature Communications* 11: 1–13. https://doi.org/10.1038/s41467-020-14496-6.

United Nations. (2018). *World Urbanization Prospects Demographic Research*. New York: United Nations.

United Nations. (2024). *World Population Prospects 2024*. New York: United Nations.

Wang, S., Liu, S., Wang, Q., Sun, Y., Yao, L. Li, D. and Meng, W. (2020). 'Dopamine modulates excitatory synaptic transmission by activating presynaptic D1-like dopamine receptors in the RA projection neurons of zebra finches'. *Frontiers in Cellular Neuroscience* 14 https://doi.org/10.3389/fncel.2020.00126.

White, C.R. and Seymour, R.S. (2003). 'Mammalian basal metabolic rate is proportional to body mass 2/3'. *PNAS* 100(7): 4046–4.049. https://doi.org/10.1073/pnas.0436428100.

Zimova, M., Hackländer, K., Good, J.M., Melo-Ferreira, J., Alves, P.C. and Mills, L.S. (2018). Function and underlying mechanisms of seasonal colour moulting in mammals and birds: what keeps them changing in a warming world?. *Biological Reviews* 93(3): 1478–1498. https://doi.org/10.1111/brv.12405.

4

POPULATION ECOLOGY

4.1 INTRODUCTION

A population is a group of organisms of the same species occupying a particular space at the same time. Populations are dynamic and change in size, location, and even existence over time. Individuals within a population interact with each other and with the environment; phenomena which are also studied by a population ecologist. Population ecology covers all aspects of population dynamics and the interactions taking place within a population.

In the first section of this chapter, a population is defined and associated difficulties that arise due to spatial and temporal factors are discussed. The focus of the following section is population structure. The main features discussed are population density, sex ratio, age structure, and distribution patterns. In the third section there is a focus on population growth. Geometric, exponential, and logistic models of population growth are described. Relevant mathematical equations used to model population growth are presented and explained. An understanding and familiarity with these equations and the ideas behind them help ecologists identify threatened species and create appropriate conservation strategies. They are also important in understanding how diseases spread and can be controlled.

The focus of the next section of this chapter life strategies which build an understanding of how species are adapted to either a rapidly changing environment or one that is relatively stable. The two life strategies considered are *r*- and *K*-selection: the first is an adaptation to a rapidly changing environment and the second to a static,

DOI: 10.4324/9781003604464-4

stable environment. Any given species will fall somewhere between the two extremes and have a mixture of *r*- and *K*- traits as are suitable for their environment.

The discussion then moves to the interactions between individuals within a population. The interactions covered include dominance, territoriality, mating systems, courtship behaviours, and parental care. Following this section there is a brief discussion of genetics and genomics.

An understanding of the dynamics of species populations allows ecologists to identify endangered species and develop targeted conservation plans. Understanding population dynamic helps assess the consequences of habitat loss, climate change, and pollution on species survival and distribution. Understanding the interactions between species and their environments, gives insights into competition, predation, and the effects of invasive species. Such knowledge can be used to identify actions necessary to maintain biodiversity and ecosystem health, and develop strategies to mitigate negative impacts and promote ecological resilience. Finally, by studying how diseases spread through plant, animal and human populations, ecologists can inform plant, animal, and public health interventions and control measures to improve health outcomes.

The focus of the case study in this chapter is applying a knowledge of population ecology to the address the current global loss of biodiversity and the development of strategies to address this phenomenon.

4.2 DEFINING A POPULATION

The definition of a population presented at the beginning of the introduction to this chapter –'a population is a group of organisms of the same species occupying a particular space at the same time' combines three elements: individuals, place, and time. While, for most species, identifying individual organisms is easy, there are some species where it is more difficult. What initially appears to a group of separate individuals is, in fact, the same individual. The world's largest living organism is thought to be Pando, a large, old, clonal colony of quaking aspen (*Populus tremula*) in Fishlake National Forest, Sevier County, Utah, USA. Pando comprises roughly 47,000 genetically identical stems (termed ramets) of quaking aspen which arise from a single, interconnected root system. As all the ramets are genetically identical, the whole colony, which covers approximately 43 hectares, is considered to be a single organism. Pando is thought to be between 16,000 and 80,000 years old. In recent decades grazing by deer and livestock, fire suppression, and limited emergence of new stems have reduced regeneration, prompting conservation efforts to protect and restore the colony. Other examples where identifying individuals can be difficult are found in bacteria, fungi, sponges, corals, hydroids, and some plants.

Defining the place in which a population resides requires setting geographical limits. This could be any scale from the global down to very small scales such as a pond or even the surface of a pond. Hence, it is possible to think of the global population of a species (say European oak (*Quercus robur*)) and of the population of that species in a small woodland. A complication in defining place can arise if two populations are sufficiently close to each other such that individuals move freely between the two (e.g. birds flying between two nearby woodlands). This is termed the 'gap-crossing problem'. Where a gap is crossed very frequently, it may be appropriate to consider the individuals in both locations as forming one population.

Defining the timescale is also sometimes complex. While some species are resident in one place for all their life cycle, individuals of other species move between different places. Examples of such movements include the daily movement of species from, say, a nesting area to a distant feeding area, the annual migration of some species of insects, birds, fish, and mammals, and the daily and seasonal movement of plankton up and down the water column in the oceans. In such instances, a population in a particular place at a particular time may range from very high to zero and the issue becomes the period of time over which the population should be defined. It is, therefore, incumbent on the ecologist to appraise themselves of the biology and behaviours of any species they are work with prior to undertaking any surveys and making ecological decisions.

4.3 POPULATION STRUCTURE

Having identified a population, it can be described and characterised according to a set of emergent properties (Figure 4.1).

Characteristics of individuals	→	**Characteristics of populations**
Age		Age (or stage) distribution
Behaviour		Population number and density
Sex		Sex ratio
Size		Spatial distribution
Stage		

Processes involving individuals	→	**Processes involving populations**
Death		Changes in age (or stage) distribution
Development		Changes in sex ratio
Emigration		Population growth (changes in numbers and density)
Feeding		Natality
Growth		Mortality
Immigration		
Reproduction		

Figure 4.1 Characteristics and processes relevant to individuals and to populations.

4.3.1 POPULATION DENSITY

Population density is the number of individuals of a species within a defined area (or, in the case of water, a defined volume) at a particular time. This simple measure is often described as the crude density as it does not take account of the structure of the landscape. To illustrate this issue in a very simple example, if 100 woodland species are found in a 100 ha landscape comprising a woodland and a lake, the crude density of woodland species is 1 individual ha^{-1}. However, if, for the sake of simplicity, 50 ha of that landscape are woodland and the other 50 ha are a lake then the population density of woodland species within the woodland is 2 individuals ha^{-1}: double the crude density. This revised density, the density of a population in the habitat that it can live in, is called the ecological density.

As population density changes, so do several environmental variables. Population density and these environmental variables are linked by feedback loops. For example, the population density of herbivores within a given area can increase until a point is reached at which there is insufficient vegetation (an environmental variable) to sustain the whole population. There is then a negative feedback loop: more individuals lead to less food per individual, and some herbivores will not eat enough to live and will die, thus reducing the population density. The classic example of this is the introduction, in 1944, of 29 reindeer (*Rangifer tarandus*) on to St. Matthew Island, Alaska, USA, a remote, uninhabited island in the Bering Sea. At the time of the introduction, there was an abundance of lichens, the reindeer's food source, and no natural predators. Over the next two decades, the population exploded to over 6,000 individuals. The reindeer consumed nearly all the lichen and other food sources, with the result that the population collapsed and by 1966, there were only 42 reindeer on the island (Klein 1968).

The example of St. Matthew Island illustrates one density-dependent factor, in this case intraspecific competition (i.e. competition between individuals of the same species, in this case for food). Other density-dependent factors are listed in Figure 4.2. There are also density-independent factors (e.g. physical and chemical environmental factors) that affect the size and density of a population. Figure 4.2 contains a more extensive list of density-independent factors).

The influences of various factors on population size are outlined in (Table 4.1).

Reproductive time lag can also affect population size. This is the lag between an environmental change and time for that change to be seen in population size. The determining factor is the length of the interbirth interval (IBI) – the time between consecutive births and comprises the gestation period (or its equivalent) and the time that the mother devotes to caring for offspring before being ready

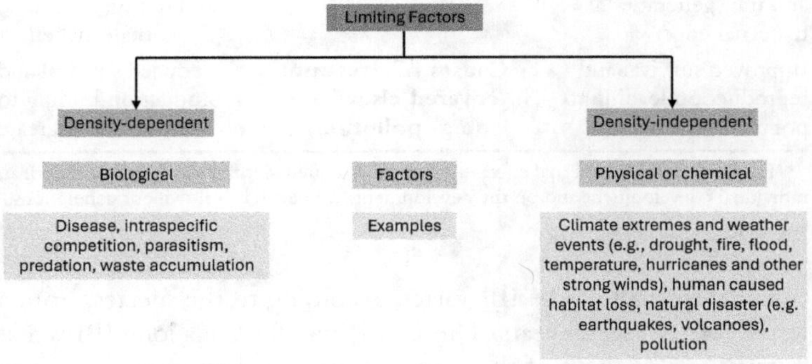

Figure 4.2 Density-dependent and density-independent factors influence on populations.

Table 4.1 Processes that influence population cycles

Decrease of the variable leads to	*Variable*	*Increase of the variable leads to*
Decreased population	**Fertility**	Increased population
Decreased population	**Survival**	Increased population
Increased reproduction and transgenerational maternal effect*	**Population density (crowding)**	Reduced reproduction and transgenerational maternal effect
Increased reproduction and transgenerational maternal effect	**Predator related stress**	Reduced reproduction and transgenerational maternal effect
Static genetic mix of population	**Dispersal among population**	Genetic mixing among populations
Static or decreasing area occupied by the population	**Dispersal to vacant habitat**	Establishment of new/ spread of existing populations
Increased reproduction and transgenerational maternal effect	**Sublethal infection**	Reduced reproduction and transgenerational maternal effect
Increased survival and population increase	**Lethal effect of predators, parasitoids, parasites, infection (can be linear, curvilinear, or threshold density-related)**	Reduced survival and population decline
Increased reproduction and transgenerational maternal effect	**Social interactions**	Reduced reproduction and transgenerational maternal effect
Improved survival and reproduction leading to population increase	**Causes of stress not covered elsewhere (e.g. pollution)**	Reduced survival and reproduction leading to population increase

* Transgenerational maternal effect – where environmental influences impact on both an individual's development and on the development and health of subsequent generations.

Source: based on Myers (2018)

to conceive again. The IBI varies, according to the species, from a few weeks to several years. The knock-on effect of a long IBI is that populations of these of these animals can be slow to recover from anything that reduces their numbers (e.g. a natural disaster, habitat loss, a disease, or poaching).

4.3.2 SEX RATIO

The ratio between males and females in a population varies through the life history of the species (due to sex/age related mortality). As a result, sex ratios are described in one of four somewhat subjective subdivisions (Table 4.2).

At birth, Fisher's principle (named after Ronald Fisher, 1890–1962, a British polymath) states that in sexually reproducing species natural selection favours a 1:1 sex ratio (Fisher 1930). Such 'Fisherian' sex ratios are commonly observed in species where males and females have similar reproductive costs and opportunities. Non-Fisherian sex ratios (i.e. where the ratio is not 1:1) occur as a result of environmental conditions, genetic mechanisms, or social structure.

In the pre-birth stage, mechanisms that can bias sex ratio at conception or at birth include various selfish genetic elements (which enhance their own transmission), microorganisms (e.g. the bacterium *Wolbachia*), and diverse sex determination systems. *Wolbachia*, a maternally inherited endosymbiotic bacteria found in many arthropods, manipulates host reproduction to promote its own transmission. For example, it alters host sex determination pathways, through mechanisms such as feminisation of genetic males, induction of parthenogenesis, and male-killing of infected embryos. Feminisation has been observed in isopods and butterflies, where *Wolbachia* causes genetic males to develop as functional females by interfering with hormonally regulated sex determination process. In some insects, infection by *Wolbachia* induces parthenogenesis. This leads to all-female progeny from unfertilised eggs. Elsewhere male killing by *Wolbachia* skews sex ratios by selectively eliminating male embryos.

Table 4.2 Definitions of different categories of sex ratios – these definitions can be somewhat subjective since they lack clear boundaries

Sex ratio category	*Definition*
Primary sex ratio	the ratio at fertilisation
Secondary sex ratio	the ratio at birth
Tertiary sex ratio	the ratio in sexually mature organism
Quaternary sex ratio	the ratio in post-reproductive organisms

Sex determination can be genetic, environmental, or social. In mammals, if the offspring has two X chromosomes (i.e. homozygous) it will, typically develop as a female, and if it has an X and a Y chromosome (i.e. heterozygous) it will be, typically develop as a male. In birds and reptiles sex determination primarily is determined by the ZW system (rather than the XY system of mammals). In birds males are typically homozygous for the Z chromosome (ZZ), and females are typically heterozygous (ZW). In reptiles the situation is the same as in birds but with a complication that environmental factors play a significant role in sex determination of some species. In amphibians the situation is not clear, as most species have homomorphic sex chromosomes (i.e. the sex chromosomes cannot be distinguished one from the other). However, it is not the chromosomes that determine the sex of an individual; it is the genes carried by those chromosomes. For example, in mammals it is the *SRY* gene on the Y chromosome that kick starts the development of male characteristics. If that gene is absent from the Y chromosome, the individual will develop with female characteristics. In some species (e.g. fruit flies (*Drosophila* spp.), zebrafish (*Danio rerio*), and chicken (*Gallus gallus domesticus*)) individual cells have their own sexual identity based on the genes they contain or express, rather than being influenced by the body's overall hormone levels. This can give rise to individuals called gynandromorphs, which exhibit both male and female characteristics (e.g. individual butterflies that have male wing colouration and patterns on one side of the body and female on the other) (Jacobs 2025).

Environmental factors that influence sex determination include temperature, pollutants (especially endocrine-disrupting chemicals), pH, aspects of the social environment, and water availability. Temperature-dependent sex determination (TDS) is observed in some species of shrimp, reptiles, teleost fish, and in a very few birds (e.g. Australian bush turkey (*Alectura lathami*) which incubates their eggs in mounds of composting vegetation. TDS is well documented in reptiles, where high temperature during development usually favours females, and fish, where high temperature consistently favours males. For temperature-induced sex reversal, unambiguous cases are limited to vertebrates, with higher temperature usually favouring males in fish and amphibians, with mixed effects

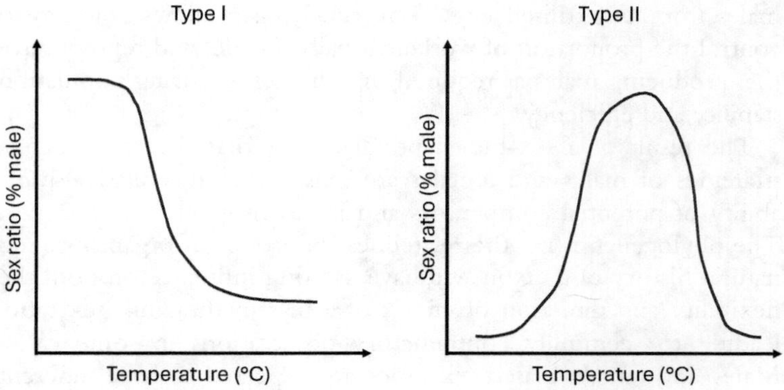

Figure 4.3 Temperature effects on the sex ratio at hatching of some reptiles.
Source: based on Shoemaker and Crews (2009)

in reptiles (Edwards 2021). Changes in local and global temperature can, therefore, affect the ratio of males and females that are born and survive (Figure 4.3).

It is not only the sex ratio of young that is affected by temperature. Sex-biased heat tolerance has been documented in a large taxonomic range including arthropods, chordates, protists, and plants. In most examples, superior heat tolerance is more common in females than males, but the direction of tolerance appears to be variable between species (Edwards 2021).

In the juvenile and subadult stage, males and females may have different sensitivities to environmental stressors, such as food and diseases. In the adult stage, differences in survival between males and females can be due to ecology, life-history, and behaviour (e.g. differences in the susceptibility of adults to predators, parasites, and diseases) (Schacht et al. 2022). In considering sex bias in populations and behaviours, it is, therefore, necessary to consider operational sex ratio (OSR), which excludes adults that are sexually inactive, pregnant, and parenting.

Socially controlled sex ratios occur when the sex of offspring is strategically adjusted based on population needs. Certain insects, such as Hymenoptera (i.e. ants, bees, and wasps) use haplodiploidy, a genetic system in which females arise from fertilised eggs and males from unfertilised eggs. This mechanism allows colonies to control the proportion of worker (usually female) and reproductive (i.e. producing males as required) individuals, ensuring population stability and efficiency.

The result of a sex-biased population is that the reproductive strategies of males and females are sensitive to the relative availability of potential competitors and mates (Kappeler et al. 2022). The phylogenetic (i.e. the relatedness of groups of organisms) and natural history of a taxon will have a strong influence on both the flexibility and direction of effects of a bias in the adult sex ratio, leading to seemingly contradictory observations in some cases. Male- and female-biased sex ratios are expected to have different consequences, with male bias expected to be more detrimental to population growth and viability, as both are often constrained by female fecundity. Skewed sex ratios may be particularly common at the edges of a species' geographical range; this affects the species' ability to extend its range as climate envelopes shift (Table 4.3).

In some species it is not possible to separate individuals as female and male; they are hermaphrodites – individuals capable of producing viable male and female gametes enabling a form of sexual reproduction in which either partner can act as the female or male. Hermaphrodism is common in plants with the same induvial having ovaries (where the female gametophyte is produced) and anthers (where the male gametophyte is produced). Hermaphroditism in animals is primarily found in the invertebrates. The great majority of tunicates, molluscs, and earthworms are hermaphrodites. Hermaphroditism is also found in some fish species but is rare in other vertebrate groups. Hermaphroditism in fish can be synchronous or sequential. Synchronous hermaphroditism occurs when one individual has both ovaries and testicles and can function as either sex at any one time. For example, black hamlets (*Hypoplectrus nigricans*) take turns releasing sperm and eggs during spawning. Sequential hermaphroditism occurs when an individual switches sex, usually from female to male (protogyny). The change is typically due to a competition for supremacy between male and females (Ortega-Recalde et al. 2020). An example of this occurs in coral reef fish such as groupers (Epinephelidae), parrotfish (Labridae), and wrasses (Labridae). When a dominant male is removed from a group of females, the largest female in the harem switches sex over

Table 4.3 Expected effects of adult sex ratio variation. ♀ = female; ♂ = male

Characteristic		*Female-biased adult sex ratio population*	*Male-biased adult sex ratio population*
Fitness (reproductive success)		Higher for ♂ than for ♀	Higher for ♀ than ♂ for
Hormone physiology (testosterone and glucocorticoids)		♂ testosterone high due to frequent ♂–♀ interactions	♂ testosterone high due to frequent ♂–♂ competition or low due to interference of testosterone with paternal behaviours ♀ glucocorticoids high due to ♂ harassment
Mate choice (choosiness)		♀ choosiness low	♂ choosiness low
Parental care and mating system (see Table 4.8)	Investment	♂ investment and care low	♂ investment and care high
	Cooperation	♂–♀ cooperation low	♀–♀ cooperation low
	Mating system	Polygyny (one male mates with more than one female)	Polyandry (one female mates with more than one male)
Sexual conflict	Intraspecific competition	♀–♀ competition/ aggression high	♂–♂ competition/ aggression high, or ♂ investment in parental care high
	Mate guarding, courtship, reproductive tactics	♂ investment in courtship low, ♀ resistance to mating low	♂ mate guarding high. Harassment and coercion of ♀s high, sneaky behaviours high, copulation duration high, ♀ resistance to mating high
Social behaviour (intrasexual interactions)		Little is known on the ASR (adult sex ratio) effects on social behaviour (e.g. coalition formation, reconciliation, vigilance)	

Source: based on Kappeler et al. (2022)

a few days and replaces the dominant male (Shapiro 1984). It is less common for a male to switch to a female (protandry). Gobies (Perciformes) can change sex in either direction depending on the sex that they interact with most (i.e. it is frequency dependent) (Schacht et al. 2022).

Sex reversal refers to individuals that have the physical features of one sex but the genetic makeup of the other. Hall et al. (2025) compared the reproductive organs of 500 birds belonging to five common Australian species to their genetic sex as indicated by their DNA. The five species in the study were the Australian magpie (*Gymnorhina tibicen*), laughing kookaburra (*Dacelo novaeguineae*), crested pigeon (*Ocyphaps lophotes*), rainbow lorikeet (*Trichoglossus moluccanus*), and the scaly breasted lorikeet (*Trichoglossus chlorolepidotus*). Sex-reversed individuals were found in all five species – at rates varying between 3 and 6 per cent. Nearly all cases were genetically female but had male reproductive organs, but there were a few genetic males with ovaries – including a genetically male laughing kookaburra with a distended oviduct, indicating it had recently laid an egg. Sex reversal indicates that determining the sex ratio of a population may not be a simple case of looking at external morphological features or at the genetics of individuals. It is not clear what the baseline rates of sex reversal are in different species and how these are changing over time. Nor is it clear if the driving forces for sex reversal are chemical (e.g. endocrine disrupters – that can be natural but are also found in products including plastics, cosmetics, pesticides, food packaging, and birth control pills), environmental stress (e.g. temperature), or some, yet to be identified factors.

4.3.3 AGE STRUCTURE

The age structure of a population is frequently shown in a life table. Life tables feed directly into demographic analyses – allowing survivorship curves to be plotted, and generation time and population growth rates to be calculated. These pieces of information can inform conservation actions (identifying life stages where mortality reduction will most increase population growth) and pest management (targeting the most fecund stage).

Population ecologists use two principal formats of life table – horizontal or cohort life tables and vertical or static – to describe how survival and reproduction vary with age or stage (Figure 4.4). To construct a horizontal or cohort life table data are collected from a group of individuals born at the same time (i.e. a cohort) from birth to death detailing how many die and how many survive in each time period. In the case of a vertical or static (time-specific) life table data are collected of the ages or stages of individuals alive at a single point in time. From this initial data collection, it is possible to calculate various parameters such as those shown in Table 4.4.

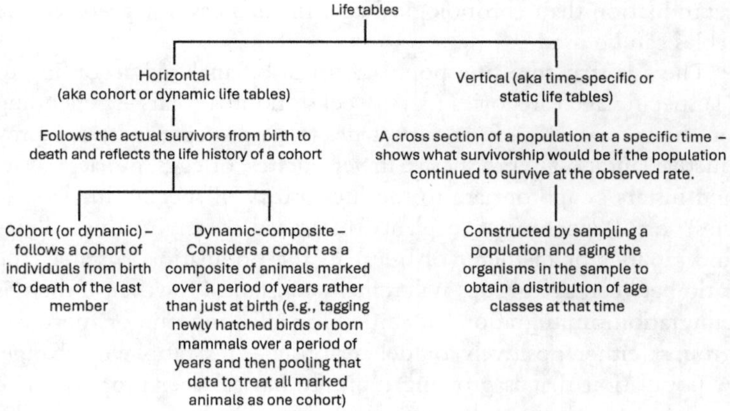

Figure 4.4 The different types of life tables.

Table 4.4 A horizontal (cohort) life table

Age in years (x)	*Number of individuals alive (N_x)*	*Proportion surviving (l_x)*	*Number dying within interval (d_x)*	*Rate of mortality (q_x)*
0	125	0.280	90	0.720
1	35	0.571	15	0.429
2	20	0.600	8	0.400
3	12	0.250	9	0.750
4	3	0.333	2	0.667
5	1	0.000	1	1.000
6	0			

Cohort tables are ideal for short-lived or easily tracked organisms (e.g. a cohort of annual plants or a brood of insects through larval, pupal and adult stages) because age-specific mortality and fecundity across a generation can be measured directly. Static tables are more practical for long-lived species (e.g. trees or large mammals) where following a full cohort is impractical. By sampling the population at a specific time and estimating age structure, it is possible to infer mortality patterns. There are situations when the size of an individual or its developmental stage (i.e. for plants seedling, juvenile, adult and for insects with complex life cycles) better predicts survival and reproduction than chronological age. In such cases stage-based life tables can be used.

The age structure of a population can be analysed according to (1) specific age categories (days, weeks, months, years – depending on the species), (2) life-history stages (e.g. pre-reproductive, reproductive, and post-reproductive in vertebrates, or eggs, pupae, larvae, and instars as appropriate to the life-history of specific insect species), or (3) size classes in plants (measured as heights of seedlings and saplings or diameter of trees). If the population is stable, the ratio between age groups will remain constant. However, if there is emigration, immigration, or something affecting one or more age groups (either positively or deleteriously), this ratio will change. A population that is growing typically has a larger proportion of younger age classes, whereas a declining population has a smaller proportion of these age classes. The age structure for a growing population tends to be pyramidal and that of a declining population tends to be column-shaped (Figure 4.5).

Biennial plants take two years to complete their life cycle e.g. foxglove (*Digitalis purpurea*). For such species, it may be appropriate to record the number of individuals in their first and second years. Recording the age of perennial plants may not be the best way to characterise a population. For example, as trees become dominant, they grow bigger than nearby individuals of the same species and age. Rather than age, the study of perennial plant populations is usually based on the size of the individual. This may be recorded as height or girth. For trees girth is commonly recorded as the diameter at breast height (abbreviated to dbh) (Figure 4.6). Breast height is defined differently in different countries, but it is usually taken to be around 1.37 m above ground level.

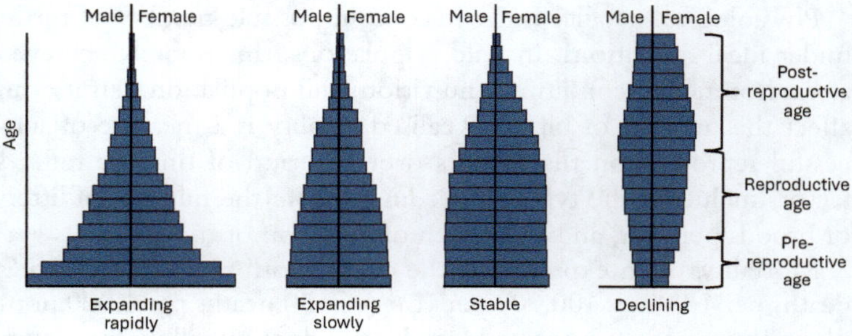

Figure 4.5 Age structures of populations that are expanding, stable, or declining.

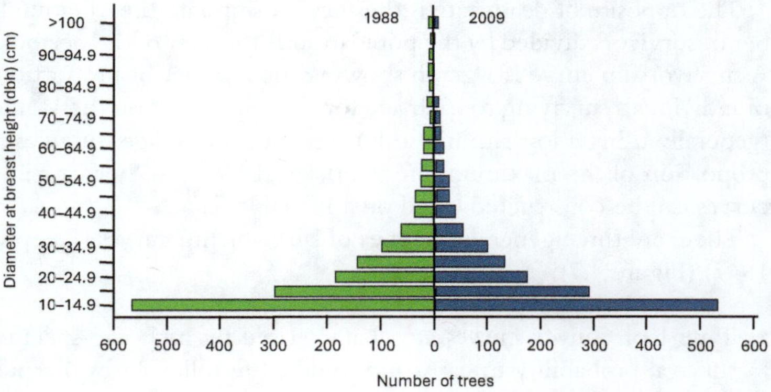

Figure 4.6 Changes in the age structure of a population of trees based on diameter at breast height (dbh).

Source: redrawn from Cascante-Marín et al. (2011)

The age structure of a population reflects the birth (natality) and death (mortality). Natality is the biggest influence on population growth. Natality, measured as a rate, can be expressed as the crude birth rate (the number of births in a given period of time (e.g. a year or a breeding season) as related to the population (e.g. 50 births per 1,000 population)), the specific birth rate (expressed relative to specific criteria such as age, or number of females), or age-specific schedule of births (the number of offspring produced per unit time by females in different age classes).

Physiological natality is the maximum possible number of births under ideal conditions. In wild populations, this is rarely achieved as environmental conditions, nutrition, and population density can affect the number of births. Realised natality is a measure of successful reproduction that occurs over a period of time. It reflects factors including the type of breeding season, the number of litters or broods per year, and the length of gestation or incubation.

Mortality, often expressed as the crude death rate is presented as deaths per 1,000 or 100,000 per year (e.g. 50 deaths per 1,000 population per year). It is termed 'crude' as it does not allow for factors that influence mortality rates such as age or health status. Mortality is also described as life expectancy and defined as the mean number of years an individual of a given age can expect to live.

The opposite of death rate is the survivorship rate (i.e. the number of survivors divided by the population at the start of the period). A survivorship curve is a graph showing the number or proportion of individuals surviving to each age for a given cohort on the *y*-axis (generally using a logarithmic scale), against time or age (often as a proportion of the maximum life span), on the *x*-axis. Survivorship curves can be constructed based on a life table data.

There are three generalised types of survivorship curves (Deevey 1947) (Figure 4.7):

- Type I or convex curves are characterised by high age-specific survival probability in early and middle life followed by a rapid decline in survival in later life (e.g. large mammals).
- Type II or diagonal curves are characterised by a roughly constant mortality rate/survival probability regardless of age (e.g. most plants, small mammals, and birds).
- Type III or concave curves are characterised by having the greatest mortality (lowest age-specific survival) early in life, with relatively low rates of death (high probability of survival) for those surviving past early life (e.g. many fish, amphibians, and invertebrates).

4.3.4 DISTRIBUTION PATTERNS

The distribution of individuals within a population across a landscape may be classified as clumped (also called clustered, contagious, or aggregated), uniform (also called regular), or random (Figure 4.8).

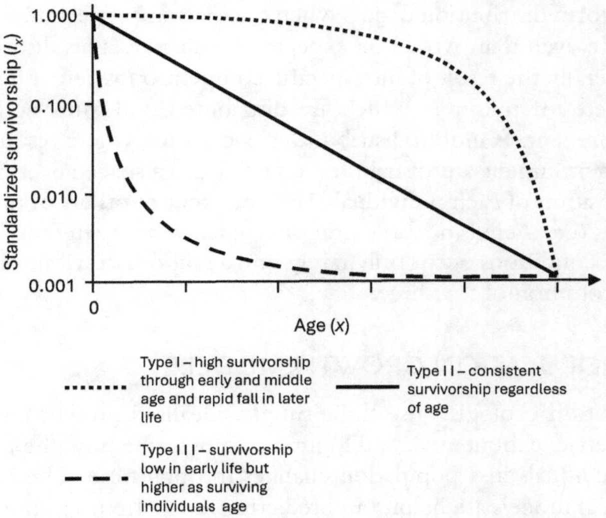

Figure 4.7 Type I, II, and II survivorship curves (note: the *y*-axis has a logarithmic scale).

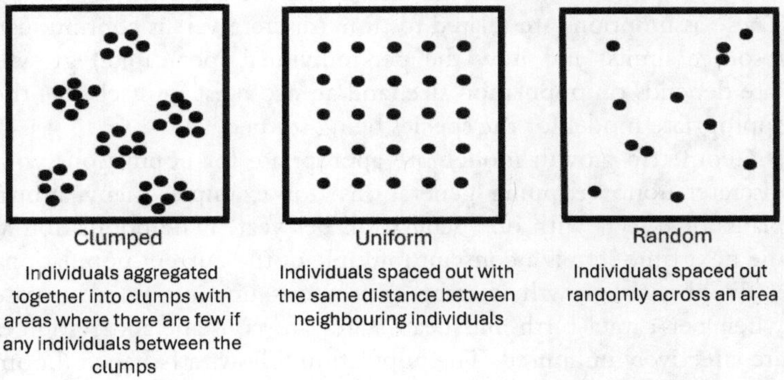

Figure 4.8 Spatial distribution patterns of individuals within populations.

The most commonly occurring distribution is clumped. It occurs where there are habitat differences across a landscape. These differences may be the result of daily or seasonal weather patterns, environmental changes, reproductive patterns, and social behaviour.

A uniform distribution occurs when the spacing between individuals is more even than would be expected by chance. This distribution is generally the result of intraspecific competition where individuals compete for resources which are distributed uniformly across the environment. Random distributions occur only where resources in the environment are distributed equally across space and time and the position of each individual is independent of others of the same species (i.e. there is no attraction or avoidance between individuals). These conditions occur only rarely, and a random distribution is the least common of the three.

4.4 POPULATION GROWTH MODELS

Population ecologists use three simple, idealised growth models – geometric, exponential, and logistic – to describe how the number of individuals in a population changes through time. These mathematical models are helpful in predicting future trends in the size of a population and recognising when there is a deviation from what might be expected. Recognising such a deviation can be a starting point for instigating a conservation programme.

The assumptions underlying each model are different (Table 4.5). These assumptions are related to time (discrete versus continuous), resource limits, and how the per-individual (per-capita) growth rate depends on population size. and an ecologist must choose the appropriate model for the species being studied.

Geometric growth models are appropriate for populations with discrete, nonoverlapping generations (for example, many annual plants or insects with one generation per year). The population at the next time step is a constant multiple of the current population.

Exponential growth describes continuous, unconstrained increase when per-capita birth and death rates are constant and resources are effectively unlimited. The population follows a J-shaped (boom and bust) curve where growth accelerates as the population grows (Figure 4.9). The population is controlled by density-independent factors, for example, in temperate climates, a population of insects may grow exponentially during the warm months of spring and summer, but when the weather cools and frosts arrive in autumn and winter, (i.e. a physical, density-independent change in the environment), the population plummets.

Table 4.5 Assumptions underlying geometric, exponential, and logistic population growth models

Attribute	*Geometric (aka discrete time model)*	*Exponential*	*Logistic*
Time frame	Discrete time steps (e.g. every month or every year). Useful for species that have synchronous reproductive cycles (e.g. breed every year or every two years e.g. seals and deer).	Continuous (or very small-time frames). Useful for species that breed continuously and do not have regular reproductive cycles (e.g. rats and mice).	Continuous (or very small-time frames). Useful for species that breed continuously and do not have regular reproductive cycles.
Ecological assumptions	Nonoverlapping generations; no density dependence.	Unlimited resources; no density dependence.	Resource limitation; density dependence via carrying capacity.
Per-capita growth rate	Constant – a constant multiplication factor can be applied at each time interval.	Constant rate – the rate of increase is constant over time.	Declines linearly as the population size approaches the carrying capacity.
Long-term growth shape curve	Geometric series; stepwise growth.	J-shaped accelerating curve.	S-shaped (sigmoid) levelling at the carrying capacity.

Exponential growth often approximates early phases of colonisation or recovery after a crash, but it is biologically unrealistic as a long-term model because it ignores density-dependent limits. The exponential growth model is particularly important in understanding disease dynamics. Contagious diseases are caused by pathogens (viruses, bacteria, fungi, and parasites). Different diseases affect plants, animals and humans (thought there is sometimes a cross over form one animal species to another). It takes time for immunity to build up to a new pathogen as the host's immune system learns to identify, attack, and kill the pathogen. This means that as a new pathogen enters a host population all, or virtually all, the individuals in the host population are susceptible to the pathogen. In these cases, the pathogen spreads exponentially. The doubling time for

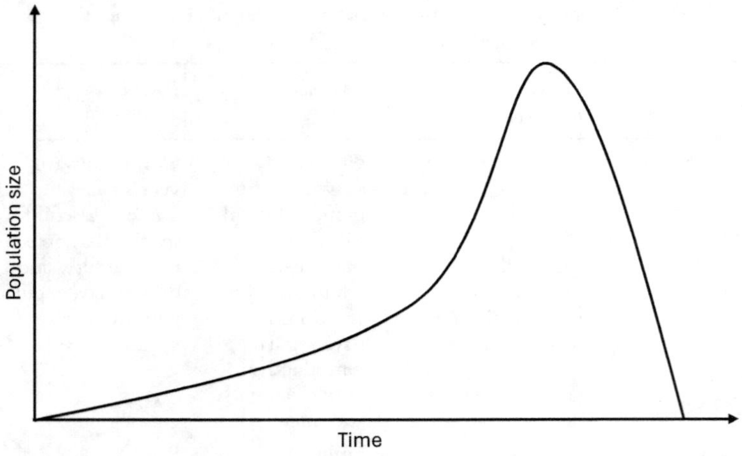

Figure 4.9 J-shaped growth curve for density-independent populations such as insects – so called 'boom and bust' as the population grows exponentially and then collapses very quickly.

the number of infected hosts will vary according to the biology of the pathogen. Another important parameter is the number of new cases directly generated from one infected individual. This is called R_0 (pronounced R nought or R zero) (Milligan and Barrett 2015). When $R_0 > 1$ it means that every infected individual will, on average, infect more than one other member of the population. This means that the number of infected individuals continues to rise exponentially at a rate governed by R_0 (the higher R_0, the greater the rise). In a population where $R_0 < 1$, each infected individual will infect (on average) fewer than one other individual, and hence the population of infected individuals will decrease.

The model of contagious disease spread presented here is very simple. It applies to new epidemics in an idealised susceptible population with random mixing. The spread of a contagious disease in real populations is more complex. In many cases aspects such as birth, death, loss of immunity, and age affect spread (Bjørnstad et al. 2020). Also, as a contagious disease spreads, so the number of susceptible hosts decreases (i.e. those that have not been infected and become immune). This is herd immunity. This means that the spread of the contagious disease slows over time. Spread can also be slowed by inoculation programmes.

Logistic growth accounts for density dependant factors by taking account of the carrying capacity, the maximum population size the environment can sustain. The logistic equation produces an S-shaped curve: near zero the population grows almost exponentially; growth slows as the size of the population approaches the carrying capacity (Figure 4.10). The logistic model is based on the assumption that resources are limited resources, and, hence, competition, or other density-dependent factors reduce per-capita growth as population size increases.

In reality, the population oscillates around the carrying capacity (Figure 4.11). The carrying capacity of an environment is not constant, it is related to environmental variables and the availability of resources, both factors vary over time (e.g. seasonally and between years). Examples of factors that influence carrying capacity include disease and predation levels, the favourability or otherwise of weather, and habitat quality. In one season or year a particular

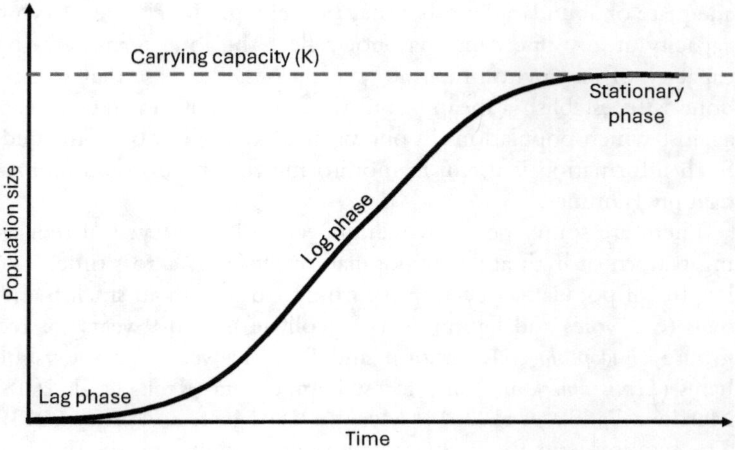

Figure 4.10 S-shaped (or sigmoid) growth curve for density-dependent populations illustrating how the growth rate of a population decreases as the population size approaches the carrying capacity of the environment.

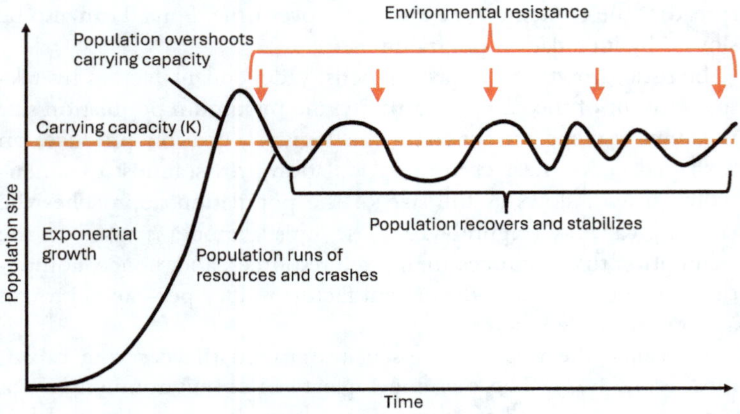

Figure 4.11 Oscillation of population size around the carrying capacity (K) for a density-dependent population.

resource may be abundant, but in the next it may be scarce and acting as a limiting factor. At any one time, while one resource may be adequate or abundant, another may be lacking. Hence, the carrying capacity at any one time may not reflect the long-term carrying capacity of an environment. In these cases, the best that can be done is to establish a mean population over many seasons or years against which populations in one particular year can be compared. Such information is useful in monitoring the success of conservation programmes.

There are some species which appear to have a cyclical repeating pattern of high and low population numbers across time. The lengths of population cycles vary from 3 to 5 years in small mammals (e.g. voles and lemmings (Arvicolinae)), 6 to 9 years for red grouse (*Lagopuslagopus scoticus*), and 8 to 11 years for snowshoe hares (*Lepus americanus*) and forest Lepidoptera (Krebs et al. 2018; Martínez-Padilla et al. 2014; Myers 2018; Myers and Cory 2013). The requirements for cyclic dynamics are a high reproductive rate (to allow the population increase), density-related mortality (to slow and stop the increase), and a prolonged negative condition (to delay the recovery of the population) (Myers and Cory 2013).

The equations used to describe geometric, exponential, and logistic population growth models are set out in Box 4.1.

BOX 4.1 POPULATION GROWTH MODEL EQUATIONS

Symbols used in the equations

t = time
$t + 1$ = time one time interval later than t
t_{double} = time for population to double in size
dt = a change in time
N = population size
N_o = initial population size
N_t = population size at time t
N_{t+1} = population size one arbitrary time-unit later that N_t
dN = rate change in the population size
b = birth rate
d = death rate
$(b - d)$ = the geometric rate of increase – the per capita rate of change in the size of the population over a discrete time interval; this rate is important in population biology and is given the symbol R
$R = (b - d)$ – geometric rate of increase
λ (lambda) = $(1 + R)$ i.e. the finite rate of increase
i = immigration rate
e (in open population growth model) = emigration rate
b_t = *per capita* births in the interval from time t to time $t + 1$
d_t = *per capita* deaths in the interval from time t to time $t + 1$.
$r = (b_t - d_t)$ – the intrinsic rate of increase
Δ = rate of chance
ln = natural logarithm
e (in exponential growth model) = a mathematical constant approximately equal to 2.71828, that is the base of the natural logarithm and exponential function
K = carrying capacity

Death and survivorship rates

$$\text{Crude Death Rate} = \left(\frac{\text{Total number of deaths}}{\text{Total population}}\right) \times 1{,}000$$

$$\text{Survivorship Rate} = \left(\frac{\text{Total number of survivors}}{\text{Total population}}\right) \times 1{,}000$$

Geometric population model

Population growth:

$$N_{t+1} = N_t + (b-d)N_t \text{OR} N_{t+1} = N_t + RN_t$$

Population at next time interval = population at previous time interval + (geometric rate of increase × population at previous time interval

Rate of change in population size:

$$\Delta Nt = RN_t$$

Change in population size = geometric rate of population growth × population

Finite rate of increase:

$$\lambda = N_{t+1} / N_t$$

Finite rate of population increase = population at next time interval ÷ population at previous time interval

Future population after t **time steps:**

$$N_t = N_0 \lambda^t$$

Population after t time steps (e.g. after 1, 2, 3, 4 or 5 years) = Initial population × finite rate of population increase raised to the power of the number of time steps

The time taken for the population to double:

$$t_{double} = \ln 2 / \ln \lambda$$

Doubling time = ln2 ÷ ln finite rate of increase

Exponential growth model

Population growth:

$$dN/dt = rN$$

Change in population size over time period t = intrinsic rate of increase × initial population size

Future population size:

$$N_t = N_0 e^{rt}$$

Population at time t = initial population × 2.72 raised to the power of the intrinsic rate of increase × t

The time taken for the population to double:

$$t_{double} = \ln 2 / r$$

Time for a population to double in size = natural logarithm of 2 ÷ the intrinsic rate of increase

Logistic growth model

Population growth:

$$dN/dt = rN((K - N)/K)$$

which can be simplified as

$$dN/dt = rN(1 - N/K)$$

Population change in time t = the intrinsic rate of increase × the population size × (carrying capacity – population size) ÷ carrying capacity

Future population:

$$N(t) = N_0 K e^{rt} / (K - N_0) + N_0 e^{rt}$$

Population at time t = initial population × carrying capacity × 2.72 raised to the power of the intrinsic rate of increase × t divided by Carrying capacity – initial population + initial population × 2.72 raised to the power of the intrinsic rate of increase × t

Population growth in an open population

Population growth:

$$r = (b + i) - (d + e)$$

Population growth rate = (births + immigration) – (deaths + emigration)

These equations are important as they are the foundations for modelling how populations change over time and can be used to identity species that are threatened and to monitor the effectiveness of species conservation programmes.

4.5 *r*- AND *K*-SELECTION STRATEGIES

In their life histories, species trade-off quantity and quality of offspring. Organisms at one extreme of a continuum of strategies invest in quick reproduction prioritising quantity over quality of offspring. This is called the *r*-selection strategy. Species at the other extreme of produce very few young and provide long periods of parental care to ensure the young are adequately nourished and protected. These species are investing in the quality, over the quantity, of offspring. This is the *K*-selection strategy. The letters *r* and *K* come from the population equations (see section 4.4).

Typically, the population size of *r*-selected is far from the carrying capacity, and thus the population grows exponentially. *r*-selected species tend to be found in abruptly changing environments where, for example, disease, predation, temperature, drought, or flooding, regularly decimate the population so that it never actually reaches the carrying capacity. In contrast, the population size of *K*-strategy species tends to remain close to the carrying capacity, where resources are limited and there is strong competition between individuals, which means that only the strongest, largest, most developed, or most intelligent members of the species survive and reproduce. Typically, *K*-strategy species are found in stable or gradually changing environments. Traits that differentiate between *r*- and *K*-strategy species are set out in Table 4.6. Some examples of *r*-strategy and *K*-strategy species are presented in Figure 4.12.

For both *r*- and *K*-strategy species there is a stable point at which population size and population growth are in balance. Additionally, there is a point for the *K*-strategy species termed

Table 4.6 Characteristics of *r*- and *K*-strategies

Trait	*r-strategy species*	*K- strategy species*
Adult body size	Small	Large
Age at first reproduction	Reproduce at an early age	Reproduce at a late age
Care for offspring	Little care	Much care
Climate	Variable and/or unpredictable; uncertain	Fairly constant and/or predictable; more certain
Competition (intra- and inter-specific)	Variable, often lax	Usually, intense
Energy efficiency	Waste a lot of energy	Energy efficient
Intelligence	Low intelligence	High intelligence
Life span	Short-lived, usually < 1 year	Long-lived, usually $>$ 1 year
Maturation rate	Fast	Slow
Mortality	Often catastrophic, non-directed, density independent	More directed, density dependent
Number of young per birth	High	Low

(*Continued*)

Table 4.6 (Continued)

Trait	*r-strategy species*	*K- strategy species*
Population size	Variable in time, non-equilibrium; usually well below carrying capacity of environment Unsaturated communities or portions thereof Ecologic vacuums, recolonised each year	Fairly constant in time, equilibrium; at or near carrying capacity of the environment Saturated communities no recolonisation necessary
Protection against predators	Not well-protected	Well-protected
Selection factors	Rapid development High maximal rate of increase r_{max} Early reproduction Small body size Single reproduction	Slower development Greater competitive ability Delayed reproduction Larger body size Repeated reproduction
Sex drive	Strong	Weak
Size at birth	Small	Large
Strength	Weak	Strong
Survivorship (see Figure 4.6)	Often Type III	Usually Types I and II

the extinction point. At this point population growth rate is lower than that required to maintain a population (Figure 4.13). In such cases, *ex-situ* (captive) breeding programmes can be instigated to increase the number of young born and the chance of survival to breeding age. In essence, a successful *ex-situ* breeding programme increases the population growth rate above that required to maintain the population size so that the population begins to grow.

In relation to plants, Grime (1977, 1979) expanded the *r* and *K* concept of life histories to include three primary plant strategies: competitive, stress-tolerant, and ruderal (Table 4.7).

Figure 4.12 Examples of *r*-strategy species (A) and (B) and *K*-strategy species (C) and (D). (A) Australian green tree frog (*Litoria caerulea*), Northeastern Australia; (B) malachite butterflies (*Siproeta stelenes*) in Serra do Caraca RPPN, Minas Gerais, Brazil; (C) Juvenile Chinese golden snub-nosed monkey (*Rhinopithecus roxellana*) in the Qinling Mountains, Shaanxi, China; (D) African bush elephant (*Loxodonta africana*), Lower Zambezi, Zambia.

Sources: (A) photograph by Brian Rogers; (B) and (C) photographs by Rob Young; (D) photograph by Luke Blazejewski

These characteristics can be illustrated graphically (Figure 4.14). Strong competitors occupy the upper part of the triangle, stress tolerators, the lower left-hand part of the triangle, and opportunistic species that are adapted to disturbance (i.e. ruderal plants), the lower right-hand part of the triangle. Species that have a mixture of traits occupy positions towards the edges and centre of the diagram.

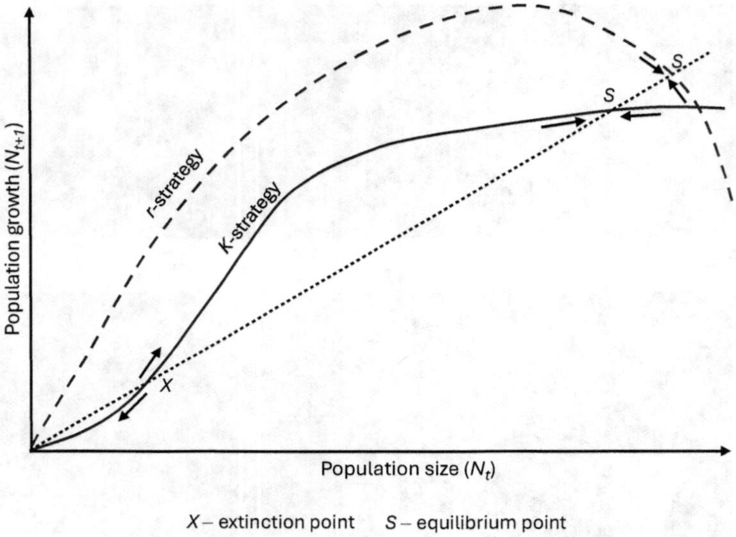

Figure 4.13 The population growth curves of *r*- and *K*-strategy species showing the stable equilibrium points (*S*) and the extinction point (*X*).
Source: based on Southwood (1976)

Table 4.7 Some traits correlated with typical competitors, stress tolerators, and ruderal plant species

Trait	*Competitive*	*Stress-tolerant*	*Ruderal*
Flowering	Annual	Intermittent	Annual
Growth rate	Rapid	Slow	Rapid
Leaf form	Variable	Leathery, needles	Variable
Leaves	Deciduous	Evergreen	Deciduous
Life form	Variable	Variable	Herbs
Litter	Persistent, copious	Persistent, sparse	Not persistent, sparse
Longevity	Long or short	Very long	Very short
Palatability to herbivores	Variable	Low	Often high

(*Continued*)

Table 4.7 (Continued)

Trait	*Competitive*	*Stress-tolerant*	*Ruderal*
Perennation (ability to survive a winter or dry season and grow again in one or more seasons)	Buds, seeds	Leaves and roots	Seeds
Reproductive effort	Small	Small	Large
Reproductive maturity	Late	Late	Early
Shoot morphology	Dense canopy	Variable	Smal stature
Stress response	Rapid	Slow	Reproduces

Source: modified from Grime (1979)

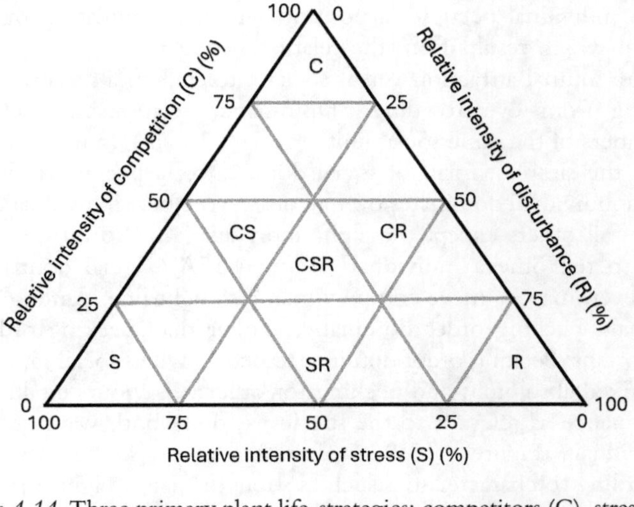

Figure 4.14 Three primary plant life-strategies: competitors (C), stress tolerators (S), and ruderal (R) plants illustrated as a CSR triangle.

Source: modified from Grime (1977)

4.6 INTRASPECIFIC COMPETITION: DOMINANCE AND TERRITORIALITY

Intraspecific competition occurs between individuals of the same species. This may be for food, space, or reproductive opportunities and can be a factor that limits population size and density within a particular habitat. Intraspecific competition manifests as agonistic (combative) behaviours, which reflect the degree of tolerance between individual of the same species. There are two forms of agonistic behaviour: dominance and territoriality. Dominance is the term used to describe behaviours that are based on the distance between individuals that provoke aggression or avoidance behaviours. Territoriality is used to describe the division and exclusive occupation of an area by an individual or social unit (e.g. a family, tribe, pod, or similar group of individuals).

4.6.1 DOMINANCE

Each individual occupies a position within a population or local group which results from the relation between the two opposing forces: mutual attraction versus social intolerance (a negative force arising from overcrowding). Dominance relationships between members of the same social unit are termed social dominance.

In the simplest form of social dominance, a pecking order, an alpha individual dominates over all others, the beta individual dominates all others except the alpha individual, and so on down the line to the omega individual (Figure 4.15A). Social dominance, however, may be more complex than a straight line. Among some animals, pecking order dominance, rather than pecking order, is important. Pecking order dominance occurs when social ranking is not fixed absolutely, and pecking (or other behaviours establishing dominance as relevant to the species) is done both ways between individuals (Figure 4.15B). Dominance may also be established according to characteristics such as strength, size, weight, maturity, previous fighting experience, previous social rank, injury, fatigue, close associations, environmental condition, and in males, the level of hormones. The dominance of an individual may vary according to changes in these characteristics. In mixed sex groups, male and female may have separate hierarchies.

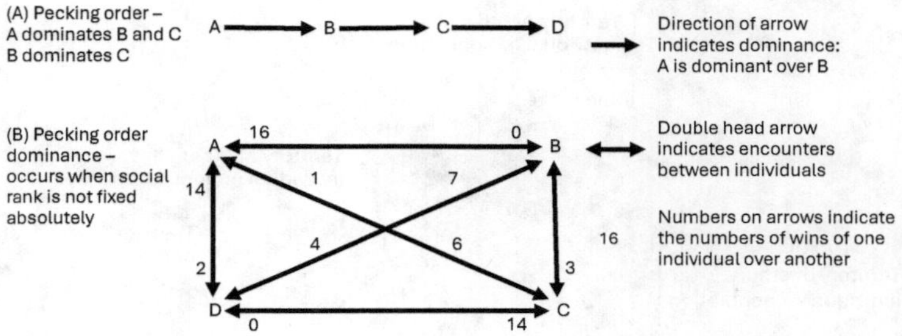

Figure 4.15 Examples of dominance structures: (A) a straight-line pecking order – A is the alpha individual, and D is the omega individual; (B) example of pecking order dominance.

Source: based on Schjelderup-Ebbe (1935)

The relationships between the two sexes vary, according to the species, from all females being dominant over all males, through the two sexes are of equal standing to all males being dominant over all females. Dominant individuals tend to get a greater share of the food available and have more mating opportunities than submissive individuals. Overall, well established/stable dominance structures reduce rates of aggression between individuals, which in turn reduces injury to and potentially death of individuals.

4.6.2 HOME RANGE AND TERRITORIALITY

A home range is the area in which an animal lives its life, seasonally, or throughout the year. The size of the home range provides ecological information on use of space by animals, resource use, predation, the distribution of organisms and populations, and their social interactions (Broekman et al. 2022). In territorial species animals defend part of their home range (i.e. their territory) either throughout the year or at certain times, for example during the time they are nesting or rearing young. Within a territory, animals also have a core area where they spend most of their time and from which they meet most of their daily needs (Figure 4.16).

Home ranges may be compact, continuous, or discontinuous, or split into two or more parts connected by trails, runways, or flight paths. An individual may not visit all its home range every day but

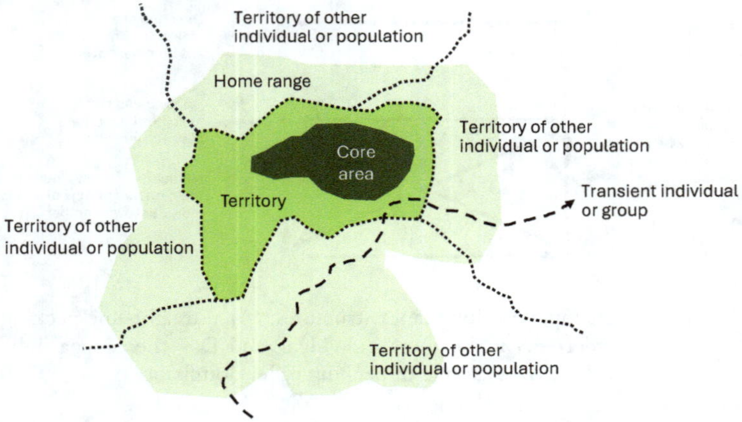

Figure 4.16 Schematic illustration of the relationship between home range, territory, and core area for an animal species.

may confine most activity to the core area (much space is used for moving between resources). Unlike territories where there is no or little overlap, home ranges frequently overlap. In the shared space, one individual is usually dominant over the other and controls the resources in that area. The subdominant individual tends to avoid the dominant individual in the shared spaces both spatially (by keeping out when the dominant individual is there) and temporally (by visiting at different times to the dominant individual).

Generally, home ranges do not have fixed or ridged boundaries and flex according to changes in the dominance of those occupying adjacent and nearby territories. For example, a pair of maned wolves (*Chrysocyon brachyurus*) living in a cerrado-Atlantic Forest transition zone in Brazil were followed using global positioning system tracking collars. The pair's showed a strong preference for rocky fields (a sub-habitat of cerrado), and a strong avoidance of Atlantic rainforest. They were associated significantly and positively with buildings and roads at night, likely because of supplemental feeding and ease of locomotion. This was the pair's 'normal' home range. During the study, the female had a litter of pups. While lactating, the female changed her home range remaining ≤250 m from water. This study showed the plasticity in home range especially of the female while lactating (Coelho et al. 2008).

Home range is a valuable and widely used metric of space use by animals. Information derived from home ranges is useful for conservation managers, such as for ensuring that protected areas are large enough to support species, identifying areas that are commonly used by multiple species, and controlling invasive species (Broekman et al. 2022).

Territoriality occurs when an individual, a breeding pair, or a social group defend a space and exclude others. This means that resources within the territory are not shared with other individuals, pairs, or social groups. Some animals hold territories only during the breeding season and live in large social groups at other times.

There are two types of territories: a general-purpose territory within which feeding and breeding take place, and a second type where nesting and breeding takes place within a territory and feeding is done elsewhere. Territories are defended in a variety of ways: song, posturing, intimidation, fighting, scent, and pheromones may all be used either in isolation or as part of a graded escalation of responses. Changes in territories occur over time. A territory holder may die or be ousted and replaced by another individual. When this happens, territorial boundaries change according to how successful the new territory holder is in defending and expanding the territorial boundaries. Within territorial species, there are often individuals which do not hold a territory (floater). These individuals may not be able to breed or may be described as sneaky maters (i.e. individuals that sneak into a territory, mate and sneak out again). A member of the floating population may take over a territory if the current holder dies, or if the floater successfully challenges the holder and forces them out. When this happens territories can be readjusted. Krebs (1971) removed six pairs of great tit (*Parus major*) from 16 breeding territories in an English oak woodland (Figure 4.17A shaded areas). Within three days, four pairs from sub-optimal hedgerow territories moved into the woodland and some of those already present expanded or contracted their territories so that the mosaic of woodland territories had no gaps – now 14 territories (Figure 4.17B shaded areas). However, the vacated hedgerow territories were not replaced suggesting that there was not a floating reserve of nonterritorial birds.

At some point in their life cycle, most plants and animals must find their own territory away from their parents. Many organisms, especially plants, rely on passive dispersal in which seeds or vegetative parts of the plant that can develop into new plants are

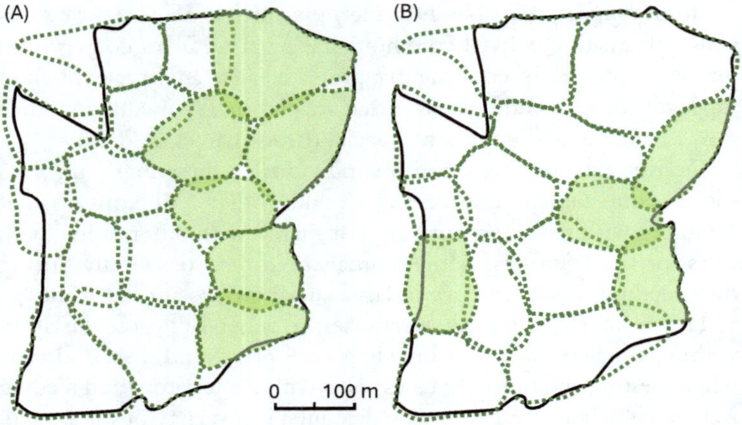

Figure 4.17 Replacement of removed great tit (*Parus major*) pairs and settlement of their vacated territories (A) territories prior to the removal of six pairs of breeding great tit (shaded territories) and (B) territories three days later by when four new pairs had established territories (shaded territories).

Source: redrawn based on Krebs (1971)

dispersed by gravity, wind, water, feathers of birds, coats of mammals, or the guts of birds and mammals. Some mobile animals rely on a passive means for transport (e.g. the wind). For example, in the marine environment the larva of sessile organisms, such as barnacles (Cirripedia), are carried by currents to new points at which they settle. A second example is that many species of spider (Arachnida), and some other invertebrates (spider mites (Tetranchidae) and several species of Lepidoptera), disperse over long distances using a technique known as 'ballooning' (sometimes called 'kiting'). To balloon the invertebrate releases several silk threads from its spinnerets into the air. These act as a parachute, which lifts the animal (Schneider et al. 2001). The Earth's static electric field may also lift ballooning invertebrates when there is no wind (Habchi and Jawed 2022).

Active animal dispersal may be undertaken by young and/or adults. It may be triggered by factors such as overcrowding, temperature change, quality and quantity of food, and photoperiod. This is especially the case in polymorphic insects, which have winged and non-winged forms, such as aphids (Aphidoidea) as leafhoppers (Cicadellidae). Young birds may disperse to find new territories after they have fledged. Adult birds, which flock together over winter months, will disperse in spring to establish breeding territories (e.g. lapwing (*Vanellus vanellus*)). Such dispersal movements drive immigration and emigration into and out of specific places.

4.7 MATING SYSTEMS, COURTSHIP, AND PARENTAL CARE

Mating strategies consist of the life-history and behaviours which organisms use to optimise fitness by allocating resources between growth, survival, and reproduction. These strategies reflect how species allocate their reproductive efforts across their lifespan. The strategies influence survival, population dynamics, and ecological roles.

There is a diversity of mating systems, including monogamy (one male with one female), polygyny (one male with more than one female), polyandry (more than one male with one female), and promiscuity (within a social group, any male mates with any female). The different systems reflect ecological constraints and parental investment strategies (see section 4.5 *r*- and *K*-selection strategies). The different strategies shape sex roles and influence sexual selection pressures which results in elaborate secondary sexual characteristics.

Courtship rituals include elaborate visual displays, acoustic signals, and chemical pheromones. These rituals ensure accurate mate selection and recognition. By regulating mate competition, resource allocation, and parental care, these behavioural frameworks govern population structure, gene flow, and adaptive capacity in changing environments. Unravelling the drivers and ecological consequences of mating strategies and courtship informs conservation initiatives aimed at safeguarding species' reproductive success.

4.7.1 ITEROPAROUS AND SEMELPAROUS SPECIES

Life history strategies can generally be categorised into two distinct types of reproduction: iteroparous (species that reproduce multiple times throughout their life) and semelparous (organisms reproduce only once in their lifetime).

The iteroparous reproductive strategy is common among many vertebrates, including mammals, birds, most fish species, reptiles, amphibians, and perennial plants. Iteroparous organisms are typically long-lived, often invest in parental care, and survival beyond each reproductive event. This approach results in a population that is stable and can adapt to changing environmental conditions. However, each reproductive event requires energy, meaning fewer offspring may be produced at one time.

Semelparous organisms invest all their available resources into a single reproductive event before dying. This strategy is found in insects, many fish species, and certain plants, such as annual flowers and bamboo (Arundinaria). The primary advantage of semelparity is the potential for producing large numbers of offspring at one time, which can be beneficial in unpredictable or extreme environments.

4.7.2 COURTSHIP

Courtship behaviours precede mating and facilitate successful reproduction. Courtship plays a critical role in species recognition, mate assessment, and synchronisation of reproductive timing. During courtship individuals recognise conspecifics (i.e. members of the same species) and evaluate the quality of potential partners. Courtship displays often function as signals of fitness, reflecting health, genetic quality, or resource-holding capacity. Courtship reduces wasted reproductive effort and maintains reproductive isolation (i.e. ensuring that only members of the same species mate with each other – if species from different species mate (e.g. a horse (*Equus ferus caballus*) and a donkey (*Equus asinus asinus*)), offspring tend to be infertile and, hence, reproductive effort is wasted (Figure 4.18).

Courtship behaviours can be classified into chemical, visual, acoustic, and tactile modes, each adapted to an animal's environment. Chemical courtship often involves the production of pheromones which are carried by air currents. Visual courtship relies on individuals being able to see each other and involves colourful displays, dances, or structural modifications such as the length of certain feathers (e.g. tail feathers). Acoustic signals range from relatively simple calls to complex songs. Tactile signals such as antennal stroking or wing vibrations provide direct assessment of partner responsiveness. These different modes can be combined into ritualised sequences of postures, vocalisations, and environmental manipulation (e.g. the building of structure to impress a potential mate) that reflect both genetic quality and learned performance.

(A) (B) (C) (D)

Figure 4.18 Courtship rituals. (A) Chinstrap penguins (*Pygoscelis antarcticus*) in Antarctica. (B) Male springbok (*Antidorcas marsupialis*) competing aggressively with each other in the Kalahari Desert, South Africa. (C) Male ostrich (*Struthio camelus*) doing a courtship display to a female in Tanzania. (D) Male ruff (*Calidris pugnax*) in a lek displaying to females in Friesland, The Netherlands.

Source: (A) photograph by Rob Young; (B) and (D) photographs by Brian Rogers; (C) photograph by Sue Melville

4.7.3 MATING SYSTEMS

Plants have complex reproductive systems. Some species are monecious (individuals carry both male and female flowers), others are dioecious (some individuals are male, and others are female), and some are hermaphroditic (an individual has flowers that produce both male and female gametes). These strategies mean that plants are variously able to self-fertilise (autogamy), cross-fertilise (outcrossing), and reproduce asexual without fertilisation (apomixis). Within one species, mixed mating systems, in which two or even all three mating systems occur, are not uncommon.

In animals, mating systems, spanning monogamy to promiscuity, determine the distribution of reproductive opportunities and the intensity of sexual selection (Table 4.8). Some species show different mating systems in different circumstances, for example, in different parts of their geographical range or under different conditions of food availability.

There is considerable variation in the reproductive strategies across taxa. The example in Table 4.9 is for vertebrates.

Table 4.8 Mating systems of animals

Mating system and sub-types	*Description and examples*
Monogamy	One male and one female have an exclusive mating relationship. The term 'pair bonding' often implies this. Monogamy is rare in the animal kingdom. Examples of monogamous animals are barn owl (*Tyto alba*), emperor penguin (*Aptenodytes forsteri*), grey wolf (*Canis lupus*), and termites (Isoptera).
Polygamy	More than one member of one sex mates with one member of the opposite sex. There are three sub-types of polygamy:
polygyny	One male has an exclusive relationship with two or more females – found in, for example, gorilla (*Gorilla* spp.), red deer (*Cervus elaphus*), elephant seal (*Mirounga* spp.), Bengal tiger (*Panthera tigris*), red-winged blackbird (*Agelaius phoeniceus*), common pheasant (*Phasianus colchicus*).
polyandry	One female has an exclusive relationship with two or more males – found in, for example, field cricket (*Gryllus bimaculatus*), honeybees (*Apis* spp.), the spider *Stegodyphus lineatus*, marmosets (Callitrichidae), and bandicoots (Peramelemorphia).
polygynandry	Two or more males have an exclusive relationship with two or more females; the numbers of males and females may not be equal. Examples include dunnock (*Prunella modularis*), European badger (*Meles meles*), and banded mongoose (*Mungos mungo*).
Promiscuity	A member of one sex within a social group mates with any member of the opposite sex. Examples include bonobo (*Pan paniscus*), red jungle fowl (*Gallus gallus*), Topi antelope (*Damaliscus lunatus jimela*), honeybee queens (*Apis* spp.), and fire-coloured beetles (Pyrochroidae).

Table 4.9 Summary of reproductive strategies across vertebrate taxa

Strategy or trait	*Fish*	*Amphibians*	*Reptiles*	*Birds*	*Mammals*
Asexual reproduction (parthenogenesis)	Yes (e.g. Amazon molly (*Poecilia formosa*))	Very rare (a few species of salamander)	Yes, in some lizards (e.g. whiptail lizards and one family of snakes)	No	No
External fertilisation	Majority especially among teleosts[1]	Majority of species	No	No	No
Internal fertilisation	Found in some fish (e.g. sharks and rays)	Found in a very small number of species (e.g. caecilians[2])	Yes	Yes, without exception	Yes, without exception
Oviparity (i.e. lays eggs)	Majority of species	Majority of species	Majority of species	Yes, without exception	Only in monotremes[3]
Viviparity (i.e. gives birth to live young)	Yes, in some species (e.g. some sharks, coelacanth, and guppy)	Caecilians; very small number of anurans and salamanders	Some snakes and lizards (e.g. common lizard (*Zootoca vivipara*)	No	Yes – all species expect the monotremes

(*Continued*)

Table 4.9 (Continued)

Strategy or trait	*Fish*	*Amphibians*	*Reptiles*	*Birds*	*Mammals*
Ovivivparity (i.e. animals give birth to live young, which have developed inside eggs within the parent's body)	Some species of shark	Some species of salamander	Yes, in some species (e.g. adder (*Vipera berus*))	No	No
Hermaphrodity (one individual can produce both male and female gametes)	Yes, widespread (especially in coral reef fish)	Occasional in frogs and salamanders	Extremely rare, resulting from abnormal development	Extremely rare, resulting from abnormal development (NB: avian bilateral gynandromorphs[4] can exist)	No

Source: based on S. Pankhurst, personal communication, 4 January 2026

Notes: [1]Teleosts – the largest group of bony fish. [2]Caecilians – limbless, worm-like amphibians belonging to the order Gymnophiona). [3]Monotreme – egg-laying mammals comprising the platypus *Ornithorhynchus anatinus*) and echidna (*Tachyglossus aculeatus* and *Zaglossus* spp.) all found in Australia or Papua New Guinea. [4]Bilateral gynandromorphs – an organism that is split laterally with one side processing female characteristics and the other having male characteristics.

Mating systems may or may not be associated with social relationships, in which the sexual partners stay together as parenting partners. Pair bonding is usual in monogamy. In many polyandrous systems, males and the female stay together to rear the young. In polygynous systems the male will often stay with one female to help rear the young, while the other females rear their young on their own. In polygynandry, each male may assist one female; if all adults help rear all the young, the system is usually termed 'communal breeding'. In highly polygynous systems, and in promiscuous systems, paternal care of young is rare or non-existent.

4.7.4 PARENTAL CARE

Parental care is a major undertaking for an animal in terms of providing energy (food), shelter, protection from predators, and may include grooming and teaching various skills required for survival. This all requires energy and time expenditure by one or both parents depending on the species. The cost of reproduction in terms of the physical drains and energy costs of brooding, gestation, birth, feeding the young, lactation, and defences of the females and young are high. Normally females invest more in terms of reproductive costs (laying eggs, or carrying young, lactation (in mammals), and parental care) and they are the sex that chooses a mate. For polygamous species, the cost of fighting off other males is so high that the life expectancy of males is greatly reduced compared with that of the females.

In the birds and placental mammals, the degree of development of the young at birth can be divided into two main classes: precocial and altricial. Precocial young can move about at or shortly after birth. The duration of parental care may be very short or may be extensive according to the species. Other species give birth to young that are altricial. These young are born helpless, naked or nearly so, and often blind and require relatively high levels of parental care. Between these two extremes are a wide range of strategies adopted by individual species (Table 4.10).

Table 4.10 Degrees of parental care in birds

Type of development	*Characteristics of the chick at hatching*				*Parents present at hatching*	*Examples*
	Down present	*Eyes open*	*Chick mobile*	*Chick can feed itself*		
Precocial 1	Yes	Yes	Yes	Yes	No	Megapods[3]
Precocial 2	Yes	Yes	Yes	Yes[1]	Yes	Ducks, Plovers
Precocial 3	Yes	Yes	Yes	Yes[2]	Yes	Quail. Turkey
Precocial 4	Yes	Yes	Yes	Yes/no	Yes	Grebes, Rails
Semi-precocial	Yes	Yes	Yes/no	No	Yes	Gulls, Terns
Semi-altricial 1	Yes	Yes	No	No	Yes	Hawks, herons
Semi-altricial 2	Yes	No	No	No	Yes	Owls
Altricial	No	No	No	No	Yes	Passerines[4]

Source: based on O'Connor (1984)

Notes:
1 Precocial 2 chicks follow parents but find their own food.
2 Precocial 3 chicks are shown food by their parents.
3 Megapods (also known as incubator birds or mound builders) are stocky, medium-large, chicken-like birds with small heads and large feet in the family Megapodiidae.
4 Passerines – also called perching birds.

4.7.5 REPRODUCTIVE VALUE

The idea that different individuals have different 'value' in terms of their contribution to future generations is called their 'reproductive value' (Fisher 1930). The reproductive value of an individual of age x is designated Vx and is the number of offspring that an individual is expected to produce over its remaining life span (after adjusting for the growth rate of the population). The reproductive value of different ages is related to an organism's life history. Typically, reproductive value is low at birth, increases to a peak near the age of first reproduction, and then declines (Caswell 2001) (Figure 4.19).

There can be differences within a species between the reproductive value of males and females. This knowledge can be useful practically in a conservation or management perspective. Knowing something about the reproductive value can suggest which individuals should be culled or transplanted.

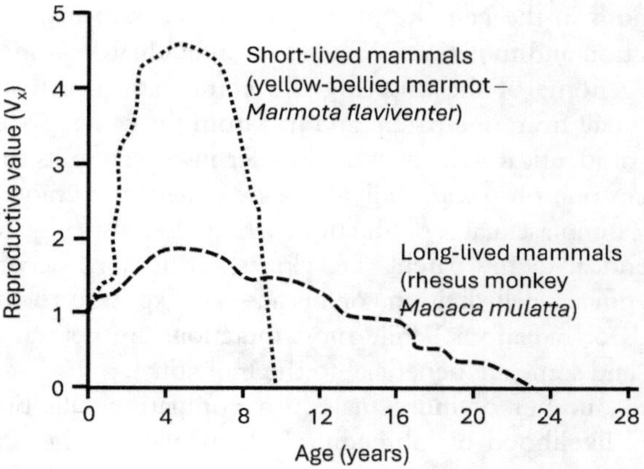

Figure 4.19 The different reproductive value of females of two species of mammals: one with a relatively rapid fall in reproductive value with age and the other with a more gradual fall-off in reproductive value with age.

Source: based on Hernández-Pacheco et al. (2013)

4.8 GENETICS AND GENOMICS

The focus of genetics is on the structure, possible variations, and function of specific genes. Genomics addresses the working of all genes (i.e. the genome – all the DNA contained in a cell including both the chromosomes within the nucleus and the DNA in the mitochondria) and their inter-relationships to identify their combined influence on the growth and development of the organism.

Genetic differences between individuals in a population can be seen in both discontinuous and continuous variations. Discontinuous variation is when the specific characteristics displayed by individuals can be separated into discrete categories. For example, the plumage of gull species (Larinae) changes as the bird matures and this allows individuals to be identified as first, second, third-year juveniles, or as adults. Continuous variation is when a character can be placed along a range of values. Examples include tail length in rodents and the number of kernels in an ear of maize (*Zea mays*). Statistically it is possible to record continuous variation in a distribution table and to plot the figures obtained in a histogram.

Variations in the genetics of individuals occurs because of sexual reproduction and mutations. When two individuals reproduce sexually the genome of the offspring is a combination of that of both parents – half from one parent and half from the other. So, except in the case of identical siblings, when the fertilised egg splits to produce more than one offspring, individuals are genetically unique. This is not the case in asexual reproduction, where the offspring are genetically identical to the parent (i.e. clones). Mutations occur spontaneously during cell division, or because of exposure to mutagenic chemicals or radiation. While most mutations are neutral some are harmful and some are beneficial to the individual.

As the number of individuals in a population falls there is an increased likelihood of inbreeding, which occurs when mates are more closely related than if the two were drawn at random from the whole population. There are consequences of inbreeding, especially if it is close inbreeding. The effects of detrimental recessive genes will be observed if offspring inherit a copy from each parent. This can lead to inbreeding depression, a term used to cover effects such as inaccurate chromosomal pairing, reduced pollen and seed production in plants, decreased fertility, decreased mating success, decreased fecundity, small body size, loss of vigour, and premature death. In captive breeding programmes it is important, therefore, to ensure that family trees (i.e. pedigrees) are recorded and that individuals are moved between populations to avoid inbreeding. It also means that when a wild population is declining, it may not be able to recover as offspring are increasingly born from closely related individuals. However, captive breeding has successfully used genetic purging (i.e. selectively removing potential recessive genes from the population by selective breeding) in the breeding of animals from a small pool of individuals such as the Przewalski horse (*Equus ferus przewalskii* or *Equus przewalskii*).

In recent years, advances in gene sequencing technologies now means that it is possible to study the whole genome. This is providing new insights into the relationships between different species, often leading to the reclassification of species. Other insights are showing relationships between distant populations of the same species, which can be helpful in conservation programmes where translocating individuals is planned, perhaps as part of a reintroduction programme.

4.9 CASE STUDY – THE BIODIVERSITY CRISIS

The beginning of the twenty-first century is marked by a period of rapid, global environmental change. Habitats are being lost and degraded, land management regimes changed (see Chapter 2), the climate is changing (see Chapter 2), fire frequency and intensity are increasing, invasive species are spreading (see Chapter 5), natural resources are being overexploited, pollution is increasing (see Chapter 2), and infectious diseases are spreading (Cullen et al. 2025). Many species cannot adapt sufficiently quickly to keep pace with the rate of environmental change (see Chapter 3). Their populations decline and they become species of conservation concern, if not extinct. For example, according to the IUCN (2024) at least 85 mammals have been driven to extinction with many more classified as globally threatened.

Extinction is a natural phenomenon. There is a natural, background rate to the loss of species: 10 per cent of species are lost every million years, 30 per cent every 10 million years, and 65 per cent every 100 million years (Raup 1991). Of the four billion species estimated to have existed on Earth over the last 3.5 billion years, 99 per cent are now extinct (Novacek 2001). Some species could not adapt to a naturally changing environment, while others evolved into new species that are distinct from the parent population which subsequently disappeared. Extinction rates have not been constant through time. There have been many periods when the extinction rate has been higher than the background rate. A mass extinction event is defined as being one in which the Earth loses more than three-quarters of its species in a geologically short interval (i.e. less than 2 million years) (Barnosky et al. 2011). In 1982 Raup and Sepkoski identified five periods of mass extinctions, though there may have been earlier mass extinction events prior to those five (Jenkins 1989). Barnosky et al. (2011) outlined these events and their possible causes (Table 4.11).

Following a mass extinction event the evolution of new species typically takes at least hundreds of thousands of years (Avise et al. 1998; Weir and Schluter 2007), and recovery from mass extinction episodes probably occurs on timescales encompassing millions of years (Alroy 2008; Lu et al. 2006). Currently, the Earth is losing biodiversity at a rate not seen since the end of the Cretaceous period 66

Table 4.11 Mass extinction events

Extinction event	*Age (mya)*	*Species lost (%)*	*Extinction trigger*	*Associate climate change*
End Ordovician	444	86	Tectonic uplift of the Appalachian Mountains resulting in weathering, sequestration of carbon dioxide leading to changes in ocean chemistry.	Intense glacial and interglacial periods, and associated sea level changes.
Late Devonian	360	75	Rapid growth and diversification of land plants.	Severe global cooling
End Permian	250	96	Volcanic activity in Siberia leading to elevated carbon dioxide and hydrogen sulphide (H_2S) levels. Leading to ocean acidification and land chemistry changes.	Global warming
End Triassic	200	80	Underwater volcanic activity in the Central Atlantic Magmatic Province (CAMP) (an area covering approximately 11 million km^2 composed mainly of basalt located in the central north Atlantic Ocean) resulting in a dramatic change in the chemical composition of the oceans.	Global warming
End Cretaceous	65	76	Asteroid impact in Yucatán, Mexico.	Rapid global cooling
Holocene or Anthropocene.	Today	Not yet known	Human actions.	Climate change, habitat loss and fragmentation, non-native invasive species

Source: based on Barnosky et al. (2011)

million years ago (Dasgupta 2021) and Barnosky et al. (2011) presented data that confirmed that the Earth is currently undergoing a sixth mass extinction known as the Holocene or Anthropocene extinction. In their review, Barnosky et al. (2011) highlight the growing evidence suggesting that humans are causing the sixth mass extinction through resource use, fragmenting habitats, introducing non-native species, spreading pathogens, killing species directly, and changing the global climate.

To address this loss of biodiversity, local, national, regional, and global transformational change across technological, political, cultural, economic, and social domains is required (IPBES 2019). Those with ecological knowledge and understanding have a key role to play in this transformational change. At one level this is to inform and sit alongside decision makers in local, national, regional, and global policy making forums (see Chapter 1 case study on evidence into policy). At another level, it is to bring about change on the ground through conservation management practices that are informed by and based on the best available ecological knowledge (see Chapters 6 and 7).

An important step in addressing the biodiversity crisis is to know what species are present and where to put conservation effort. Collecting observational data on the presences of species in a particular habitat can be costly, time consuming, and dangerous. Data collection techniques are also changing. One such relatively new technique is collecting and analysing eDNA (environmental DNA). This is a non-invasive, highly sensitive method for identifying species in various ecosystems without the need for direct observation or physical capture. Organisms shed DNA into their surroundings (water, soil, or air). The shed DNA is termed eDNA. This eDNA can be collected in a water, soil, or air sample and analysed. This approach is particularly valuable for studying elusive, rare, or cryptic species, and hard to observe environments (e.g. the deep oceans and dense forests). Other data collection techniques that are being increasingly used include camera traps, audio recording (often linked to AI), and robots (currently mainly used in aquatic environments). Collectively, these techniques are providing more data than ever on the presence and distribution of populations.

Table 4.12 Species richness in various regions in Brazil (figures in brackets are the percentage of the total flora or fauna in each category). 'Threatened' consists of all non-endemic and endemic species categorised as endangered in the IUCN Red List (2024). 'Widespread' includes all non-endemic and non-endangered native species

	Region of Brazil						
	Amazon	*Atlantic Forest*	*Caatinga*	*Cerrado*	*Pampa*	*Pantanal*	*Total*
Flora							
Total	933	2960	2756	577	693	1424	9343
Widespread	138 (14.79)	2046 (69.12)	2292 (83.16)	362 (62.74)	42 (6.06)	1088 (76.40)	5969 (63.89)
Endemic	751 (80.49)	816 (27.57)	391 (14.19)	155 (26.86)	635 (91.63)	98 (6.88)	2846 (63.89)
Endangered	49 (5.25)	202 (6.82)	103 (3.74)	125 (21.66)	159 (22.94)	326 (22.89)	964 (10.32)
Fauna							
Total	713	668	1322	93	91	335	3222
Widespread	138 (19.35)	149 (22.31)	1289 (97.50)	50 (53.76)	19 (20.88)	314 (93.73)	2361 (73.28)
Endemic	108 (15.15)	443 (66.32)	9 (0.68)	29 (31.18)	2 (2.20)	1 (0.30)	592 (18.37)
Endangered	77 (10.80)	225 (33.68)	28 (2.12)	22 (23.66)	72 (79.12)	20 (5.97)	363 (11.27)

Source: based on Liévano-Latorre et al. (2025)

Patterns of biodiversity loss are varied according to habitat and global location. Working across Brazil, Liévano-Latorre et al. (2025) found different patterns in the total number of species, the proportion that are widespread, the proportion that are endemic, and the proportion of the total number of species that are endangered (Table 4.12). This illustrates the complexity in understating what is happening to biodiversity and where change is occurring.

In 2024, Langhammer et al. published a global meta-analysis of initially identified 1445 papers published between 1890 and 2019, spanning spatial scales from local to continental. Within these, 186 studies contained temporal data from which outcomes expressed as effect sizes generated from the 'rate of change under the intervention' compared with 'rate of change under a counterfactual scenario' could be expressed. In the same paper, Langhammer et al. evaluated the impact of seven intervention types of conservation actions that address direct pressures on biodiversity, promote restoration and recovery of populations and habitat, and aim to safeguard the environment across different levels of biological organisation and compared the outcomes with those expected without intervention. The seven intervention types aiming to tackle the direct drivers of environmental degradation were:

(i) establishment and management of protected areas,
(ii) other measures to reduce habitat loss and degradation such as policy and restoration,
(iii) sustainable use of species,
(iv) sustainable management of ecosystems,
(v) control of pollution,
(vi) eradication and control of invasive alien (and problematic native) species, and
(vii) climate change adaptation.

The impacts of each intervention on biodiversity at the ecosystem, species, and genetic level were considered.

Langhammer et al. (2024) showed that in two-thirds of cases, across the full suite of conservation actions and intervention types, multiple levels and metrics of biodiversity, and over a century of action, conservation either improved the state of biodiversity or at least slowed declines compared with no conservation action. Among the different conservation actions evaluated, the eradication, control, and management of invasive and problematic species showed the largest impact of conservation action (see Chapter 5), followed by actions to reduce habitat loss and degradation (see Chapter 7), sustainable management of ecosystems (see Chapter 6), and protected areas. This meta-analysis provides strong evidence that conservation actions are successful. However, to fully address the biodiversity crisis they require transformational scaling up to meet global targets.

4.10 SELF-STUDY SUGGESTIONS

4.10.1 SELF-STUDY QUESTIONS

1. What is a population?
2. What is the significance of sex ratio to a population?
3. What factors can affect the sex ratio of a population?
4. What is the significance of age structure in a population?
5. What is an age pyramid and what information does it give about a population?
6. What are some of the problems in applying the concept of age distribution to a population of plants?
7. What are the differences between Type I, II, and III survivorship curves? Give examples of species relevant to each type.
8. What are the three basic types of spatial distribution of populations? Which is the most common and why?
9. Distinguish between geometric and exponential models of population growth.
10. Distinguish between exponential and logistic growth and give an equation for each.
11. What is the difference between lambda (λ), R_0, and r?
12. What is carrying capacity (K)?
13. Why do populations fluctuate around the estimated value of K?
14. What are the differences between density-dependent and density-independent regulation of population size?
15. What are the differences between r- and K-strategies?
16. What are the three basic type of dispersal movement and how do they differ?
17. What is the difference between dispersal and migration?
18. What is the meaning of competition in an ecological sense?
19. What are the differences between dominance, territoriality?
20. What is the home range of a species?
21. What is the purpose of courtship?
22. What are the main types of mating systems and how do they differ?

4.10.2 A DEEPER DIVE

A. How are genomics revolutionising the study of populations and their conservation?

B. How is a knowledge of population ecology used to understand how contagious diseases spread. Examples of contagious diseases include measles, pertussis (whooping cough), chicken pox, SARS2-COVID, mumps, norovirus, tuberculosis, Ebola virus disease, hepatitis A, and influenza.

C. Search the internet for examples of ex-situ (captive breeding) conservation programmes (e.g. Californian condor (*Gymnogyps californianus*).

D. Search the internet for examples of courtship behaviours in invertebrates and vertebrates.

E. How is a knowledge of population ecology used to understand and forecast growth of the human population and the potential impacts of this growth?

F. Explore the data sets available from the United Nations at https://population.un.org/dataportal/home?df=88b7d752-537c-46b4-821c-eae29af82531

4.11 REFERENCES

Alroy, J. (2008). 'Dynamics of origination and extinction in the marine fossil record'. *PNAS USA* 105: 11536–11542.

Avise, J.C., Walker, D. and Johns, G.C. (1998). 'Speciation durations and Pleistocene effects on vertebrate phylogeography'. *Proceedings of the Royal Society B* 265(1407): 1707–1712. https://doi.org/10.1098/rspb.1998.0492.

Barnosky, A.D., et al. (2011). 'Has the Earth's sixth mass extinction already arrived?'. *Nature* 471(7336), 51–57. https://doi.org/10.1038/nature09678.

Bjørnstad, O.N., Shea, K., Krzywinski, M. and Altman, N. (2020). 'The SEIRS model for infectious disease dynamics'. *Nature Methods* 17: 557–558. https://doi.org/10.1038/s41592-020-0856-2.

Broekman, M.J.E., Hoeks, S. Freriks, R. Langendoen, M.M., Runge, K.M., Savenco, E., ter Harmsel, R. Huijbregts, M.A.J. and Tucker, M.A. (2022). 'HomeRange: a global database of mammalian home ranges'. *Global Ecology and Biogeography* 32(2): 198–205. https://doi.org/10.1111/geb.13625.

Cascante-Marín, A., Victor Meza-Picado, V. and Armando Estrada-Chavarría, A. (2011). 'Tree turnover in a premontane neotropical forest (1998–2009) in Costa Rica'. *Plant Ecology* 212: 1101–1108. https://doi.org/10.1007/s11258-010-9890-y.

Caswell, H. (2001). *Matrix Population Models* (2nd edition). Sunderland, MA: Sinauer Associates.

Coelho, C.M., de Melo, L.F.B., Sábato, M.A.L., Magni, E.M.V., Hirsch, A. and, Young, R.J. (2008). 'Habitat use by wild maned wolves (*Chrysocyon brachyurus*) in a transition zone environment'. *Journal of Mammalogy* 89(1): 97–104. https://doi.org/10.1644/06-MAMM-A-383.1.

Cullen, D., Kingsford, R.T., Bino, G., West, R., Letnic, M. and Pedler, R. (2025). 'Bucking the trend – recovery from near continent-wide extinction by a marsupial micro-predator during drought'. *Biological Conservation* 311: 111411. https://doi.org/10.1016/j.biocon.2025.111411.

Dasgupta, P. (2021). *The Economics of Biodiversity: The Dasgupta Review*. London: HM Treasury.

Deevey, E.S. (1947). 'Life Tables for natural population of animals'. *The Quarterly Review of Biology* 22(4): 283–314. https://doi.org/10.1086/395888.

Edwards, S. (2021). 'Sex ratios in a warming world: thermal effects on sex-biased survival, sex determination, and sex reversal'. *Journal of Heredity* 112(2): 155–164. https://doi.org/10.1093/jhered/esab006.

Fisher, R.A. (1930). 'Sexual reproduction and sexual selection: natural selection and the sex-ratio'. In R.A. Fisher, *The Genetical Theory of Natural Selection, 141*. Oxford: Clarendon Press.

Grime, J.P. (1977). 'Evidence for three primary strategies in plants and its relevance to ecological and evolutionary theory'. *The American Naturalist* 111: 1169–1194.

Grime, J.P. (1979). *Plant Strategies and Vegetation Processes*. New York: Wiley.

Habchi, C. and Jawed, M. (2022). 'Ballooning in spiders using multiple silk threads'. *Physical Review E*. 105(3-1): 034401. https://doi.org/10.1103/PhysRevE.105.034401.

Hall, C.A., Conroy, G., Jelocnik, M., Kasimov, V., Gillet, A., Portas, T., Hill, A. and Potvin, D.A. (2025). 'Prevalence and implications of sex reversal in free-living birds'. *Biology Letters* 21: 20250182. https://doi.org/10.1098/rsbl.2025.0182.

Hernández-Pacheco, R., Rawlins, R.G., Kessler, M.J., Williams, L.E., Ruiz-Maldonado, T.M., González-Martínez, J., Ruiz-Lambides, A.V. and Sabat, A.M. (2013). 'Demographic variability and density-dependent dynamics of a free-ranging rhesus macaque population'. *American Journal of Primatology* 75: 1152–1164. https://doi.org/10.1002/ajp.22177.

IPBES. (2019). *Global Assessment Report on Biodiversity and Ecosystem Services of the Intergovernmental Science-Policy Platform on Biodiversity and Ecosystem Services*. Bonn, Germany: IPBES Secretariat. https://doi.org/10.5281/zenodo.3831673.

IUCN. (2024). 'The IUCN red list of threatened species: version 2024-1'. Retrieved from www.iucnredlist.org.

Jacobs, P. (2025). '"Sex reversal" is surprisingly common in birds, new study suggests'. *Science* 12 August. https://doi.org/10.1126/science.zkfxy3w.

Jenkins R.J.F. (1989). 'The supposed terminal Precambrian extinction event in relation to the Cnidaria'. *Memoirs of the Association of Australasian Palaeontologists* 8: 307–317.

Kappeler, P.M., et al. (2022). 'Sex roles and sex ratios in animals'. *Biological Reviews* 98(2) 462–480. https://doi.org/10.1111/brv.12915.

Klein, D.R. (1968). 'The introduction, increase, and crash of reindeer on St. Matthew Island'. *The Journal of Wildlife Management* 32(2): 350–367. https://doi.org/10.2307/3798981.

Krebs, C.J., Boonstra, R. and Boutin, S. (2018). 'Using experimentation to understand the 10-year snowshoe hare cycle in the boreal forest of North America'. *Journal of Animal Ecology* 87: 87–100. https://doi.org/10.1111/1365-2656.12720.

Krebs, J.R. (1971). 'Territory and breeding density in the great tit, *Parus major* L.' *Ecological Society of America* 52(1): 2–22. https://doi.org/10.2307/1934734.

Langhammer, P.F., et al. (2024). 'The positive impact of conservation action'. *Science* 384: 453–458. https://doi.org/10.1126/science.adj6598.

Liévano-Latorre, L.F., et al. (2025). 'Addressing the urgent climate and biodiversity crisis through strategic ecosystem restoration in Brazil'. *Biological Conservation* 302: 110972. https://doi.org/10.1016/j.biocon.2025.110972.

Lu, P.J., Yogo, M. and Marshall, C.R. (2006). 'Phanerozoic marine biodiversity dynamics in light of the incompleteness of the fossil record'. *PNAS USA* 103(8): 2736–2739. https://doi.org/10.1073/pnas.0511083103.

Martínez-Padilla, J., Redpath, S.M., Zeineddine, M. and Mougeot, F. (2014). 'Insights into population ecology from long-term studies of red grouse *Lagopuslagopus scoticus*'. *Journal of Animal Ecology* 83: 85–98. https://doi.org/10.1111/1365-2656.12098.

Milligan, G.N. and Barrett, A.D. (2015). *Vaccinology: An Essential Guide.* Chichester: Wiley Blackwell.

Myers, J.H. (2018). 'Population cycles: generalities, exceptions and remaining mysteries'. *Proceedings of the Royal Society B* 285: 20172841. https://doi.org/10.1098/rspb.2017.2841.

Myers, J.H. and Cory, J.S. (2013). 'Population cycles in forest Lepidoptera revisited'. *Annual Review of Ecology, Evolution, and Systematics* 44: 565–592. https://doi.org/10.1146/annurev-ecolsys-110512-135858.

Novacek, M.J. (2001). *The Biodiversity Crisis: Losing What Counts.* New York: The New Press.

O'Connor, R.J. (1984). *The Growth and Development of Birds.* Chichester: Wiley.

Ortega-Recalde, O., Goikoetxea, A., Hore, T.A., Todd, E.V. and Gemmell, N.J. (2020). 'The genetics and epigenetics of sex change in fish'. *Annual Review of Animal Biosciences* 8: 47–69. https://doi.org/10.1146/annurev-animal-021419-083634.

Raup, D.M. (1991). 'A kill curve for Phanerozoic marine species'. *Paleobiology* 17: 37–48. www.jstor.org/stable/2400988.

Raup, D.M. and Sepkoski, J.J. (1982). 'Mass extinctions in the marine fossil record'. *Science* 215(4539): 1501–1503. https://doi.org/10.1126/science.215.4539.1501.

Schacht, R., et al. (2022). 'Adult sex ratios: causes of variation and implications for animal and human societies'. *Communications Biology* 19(5): 1273. https://doi.org/10.1038/s42003-022-04223-w.

Schjelderup-Ebbe, T. (1935). *Social Behaviour of Birds: A Handbook of Social Psychology*. Worcester, MA: Clark University Press.

Schneider, J.M., Roos, J., Lubin, Y. and Henschel, J.R. (2001). 'Dispersal of *Stegodyphus dumicola* (Araneae, Eresidae): they do balloon after all!'. *Journal of Arachnology* 29(1): 114–116. https://doi.org/10.1636/0161-8202(2001)029[0114:DOSDAE]2.0.CO;2.

Shapiro, D.Y. (1984). 'Sex reversal and sociodemographics processes in coral reef fishes'. In G.W. Potts and R.K. Wootoon (eds), *Fish Reproduction: Strategies and Tactics*, 103–116. London: Academic Press.

Shoemaker, C.M. and Crews, D. (2009).'Analyzing the coordinated gene network underlying temperature-dependent sex determination in reptiles'. *Seminars in Cell & Developmental Biology* 20(3): 293–303. https://doi.org/10.1016/j.semcdb.2008.10.010.

Southwood, T.R.E. (1976). 'Bionomic strategies and population parameters'. In R.M. May (ed.), *Theoretical Ecology, Principles and Applications*. Oxford: Blackwell.

Weir, J.T. and Schluter, D. (2007). 'The latitudinal gradient in recent speciation and extinction rates of birds and mammals'. *Science* 315(5818): 1574–1576.

5

COMMUNITY ECOLOGY

5.1 INTRODUCTION

In ecology, a community refers to the populations of organisms (bacteria, fungi, plants, and animals) that live and interact within a specific area at a specific time. Community ecologists examine how the different species interact with each other within their shared environment (e.g. inter-species competition, predation, and mutualism).

In the first section of this chapter the concept of an ecological community is defined and discussed. The focus in the next two sections of this chapter is on, firstly, the physical and then the biological structure of an ecological community. Physical structure includes the vertical layering of vegetation and the horizontal distribution of species. The biological structure of an ecological community is characterised by species diversity and abundance, the roles species play (i.e. their niche), and by the position of species in different trophic levels.

In the following two sections the focus is on, respectively, community dynamics and interactions. Communities change over time as they undergo a process of succession. Primary and secondary succession, and the associated concept of a sere are discussed. The various ways in which species interact with each other in a community are outlined. The stability of food webs and an ecosystem's ability to withstand disturbances are both dependent on there being a rich biodiversity within a community and multiple, complex and dynamic interactions between these species. The loss of a species can have a cascading effect on the entire community. Towards the

DOI: 10.4324/9781003604464-5

end of the chapter there is a discussion of communities as discrete objects or as a spatial continuum along which community composition changes.

By studying community dynamics, ecologists seek to understand how natural systems function and respond to changes, such as those caused by a changing climate or habitat loss. Understanding community dynamics is essential for developing effective conservation strategies. For example, by identifying keystone species – organisms that have a disproportionately large effect on their environment – allows conservation programmes to be focussed on protecting these critical species and, hence, the whole system. Where funding for conservation programmes is limited, such priority identification is essential.

The case study in this chapter focusses on invasive non-native species (INNS). Such species enter a community through natural spread, or as a result of accidental or deliberate release. Once present, they outcompete extant community members and change community structure and properties. Where, as a result of INNS or other causes, natural systems have been disrupted and communities simplified (i.e. containing fewer species than in an unaffected system) community ecologists can offer a framework for restoration based on their analysis of the original community composition and structure. Plans for the reintroduction of native species, removal of INNS, and restoration of natural habitats can be drawn up.

5.2 DEFINITION OF AN ECOLOGICAL COMMUNITY

As stated in the introduction, in ecology, a community refers to the populations of organisms (bacteria, fungi, plants, and animals – viruses, though strictly non-living, can be added to this list) that live and interact within a specific area at a specific time. One of the issues that arises from this definition is that there is no clear, objective spatial boundary to a community. This means that it is possible to talk about the community of organisms associated with habitats of any size, for example, deep sea vents, coral reefs, dry grasslands, a mangrove swamp, an alpine pasture, or of a specific tree. Identification of what constitutes a community may be further complicated by seasonal changes when the species present in the community changes throughout the year as migratory species come and go.

The word community is sometimes used in a different context. It is used by some to describe a group of organisms that exploit the same environmental resource (e.g. a woodland, grassland, pond, or individual tree). The collection of large carnivores in a savannah or of the different sap-sucking invertebrates on a plant could both be described as a community. The word guild, rather than community, is preferred for such situations. A guild can be defined as a group of species that exploit the same class of environmental resource in a similar way.

5.3 PHYSICAL STRUCTURE OF AN ECOLOGICAL COMMUNITY

The physical structure of an ecological community is the spatial arrangement and physical organisation of organisms within a given habitat. This includes the vertical layering of vegetation (such as ground cover, understory, and canopy in a forest), the distribution of species across horizontal space, and the presence of physical features including rocks, water bodies, and fallen trees. These physical patterns are dynamic and are influenced by abiotic factors including climate and soil composition (see Chapter 2) as well as biotic factors such as interspecies interactions (see Chapter 4). Differences in the physical structure of a community create microenvironments that support diverse life forms, enabling coexistence and niche differentiation (i.e. the process by which competing species utilise different resources or habitats, thereby, reducing competition and allowing for their coexistence).

In aquatic systems, structure can be defined by water depth, current flow, and substrate type. For convenience the vertical structure of a water body is divided into zones. Generally, the communities in each zone differ from each other. The photic zone (from the Greek *phōs* – 'light') is the upper layer of a water column in which there is some light. This zone is dominated by autotrophic phytoplankton (i.e. microscopic plankton that can photosynthesise). The photic zone is frequently divided into the littoral zone – where the land meets the water, including the shore and shallow waters – and the limnetic zone – the open and well-lit area of a pond or lake. The key difference between these two zones is the presence of plants rooted in the substrate of the littoral zone. Due to a lack of sunlight, photosynthesis cannot take place and no such plants grow in the substrate in the limnetic zone. In extremely shallow water bodies, light may penetrate across the entire lakebed. In this situation, there is no limnetic zone, and the littoral zone spans the entire lake.

Below the photic zone is the aphotic zone (i.e. the zone where there is no sunlight). The aphotic zone can be divided into the profundal and benthic zones. The profundal zone is located below the range of effective light penetration and tends to have a low oxygen concentration. The benthic zone (from the Greek *batho* 'depth') is the lower layer and includes the sediment surface and some subsurface structures such as caves (Figure 5.1).

In the open oceans, where water depths are up to nearly 11,000 m (the Mariana Trench in the western pacific) other structural divisions can be recognised (Figure 5.2).

The vertical structure of terrestrial ecological communities often determines light availability, temperature variation, and moisture levels. All of which influence the organisms that can thrive at each level. The number of layers depends on the habitat. For example, grasslands generally have root (aka rhizosphere), ground (aka soil, moss, or cryptogam layer), and herb layers and sometimes a shrub layer. In comparison tropical and temperate forests have root, ground, herb, shrub, understory, canopy, and emergent layers (Figure 5.3).

Some layers can be absent due to natural phenomena or to human actions. An example of a natural phenomenon leading to fewer woodland layers can be found in boreal (needleleaf forests) where

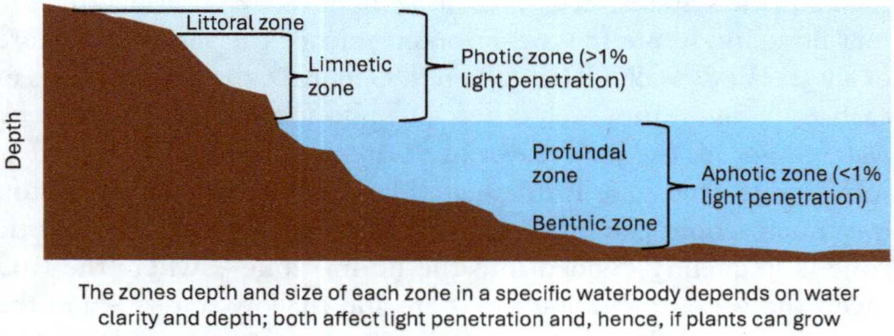

Figure 5.1 Vertical structure of a pond or lake and the associated ecological zones.

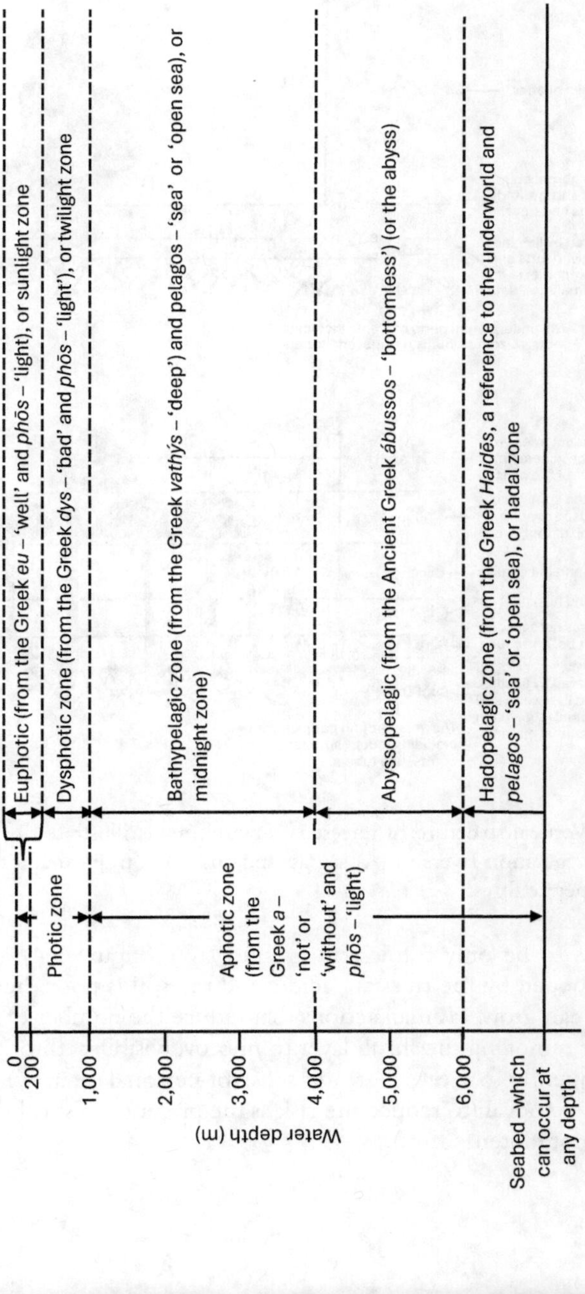

Figure 5.2 Divisions in the vertical structure of the deep oceans.

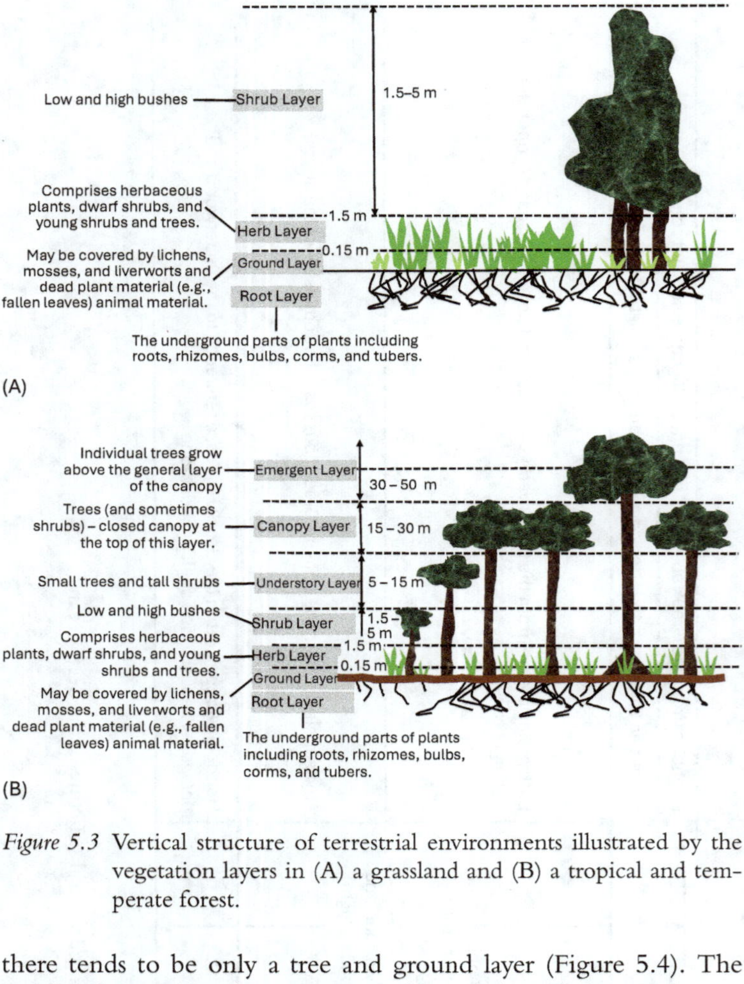

Figure 5.3 Vertical structure of terrestrial environments illustrated by the vegetation layers in (A) a grassland and (B) a tropical and temperate forest.

there tends to be only a tree and ground layer (Figure 5.4). The needles dropped by the trees are acidic and the soil is poor; hence, few plants can grow. Human actions that reduce the number of layers include removing the shrub layer to improve visibility through a forest (frequently to create a greater sense of personal security) and understory removal (to reduce fire risk as the presence of shrubs can promote more intense fires).

Figure 5.4 Vertical structure of a boreal forest with ground, herb, and canopy layers, Finland.

Source: photograph by Elmeri James Vainikainen Dennis

In both aquatic and terrestrial environments, each vertical layer is typically occupied by characteristic organisms though there can be considerable movement of individuals between layers. Such movements reflect daily and seasonal variations in the physical environment such as humidity, temperature, light, and oxygen content (of water), and shifts in the abundance of resources including food, shelter, and other requirements for an organism to complete its life cycle.

As applied to aquatic environments horizontal structure is related to differences in the material in the bed of a water body causing difference in water flow, oxygen levels, and vegetation growth (Figure 5.5).

In terrestrial environments, horizontal structure refers to difference in sizes and ages of plants, richness of plant species in the different layers, and the presence of abiotic features such as rocks or water bodies (Figure 5.6).

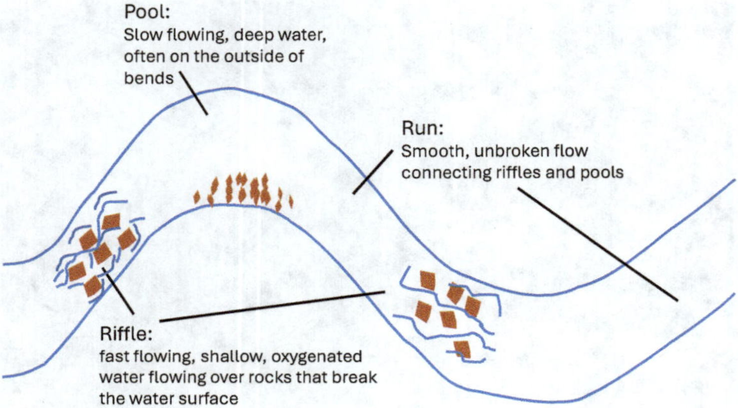

Figure 5.5 Simplified example of horizontal structure in a river or stream.

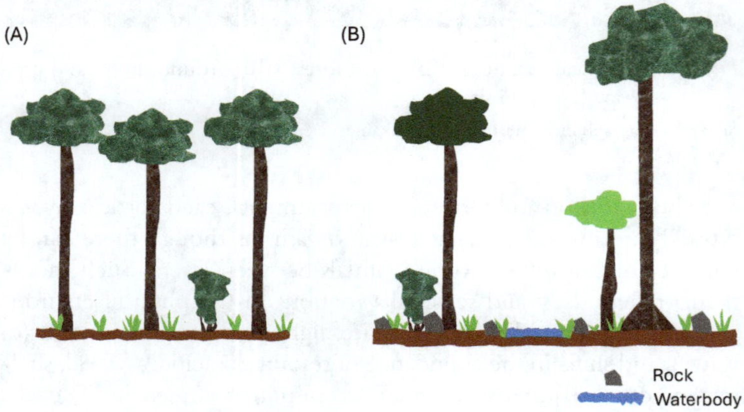

Figure 5.6 Horizontal diversity within a forest: (A) little and (B) more horizontal diversity.

Figure 5.7 contains images of contrasting forests from Western Australia illustrating variations in both vertical and horizontal structure, demonstrating that there is a continuum from low to high diversity rather than these two extremes being absolutes.

(A) (B)

Figure 5.7 Contrasting forests from Western Australia illustrating variations in both vertical and horizontal structure. (A) A forest with little vertical and horizontal diversity – the Boranup Forest between Margaret River and Augusta, Western Australia, it is the westernmost spread of Karri (eucalyptus) trees in Western Australia and a relatively young forest (100 years). (B) A forest with high vertical and horizontal diversity – the Milyeannup State Forest (part of the Jarrah Forest Biogeographic Region, Western Australia).

Source: photographs by Mike Allen

5.4 BIOLOGICAL STRUCTURE OF AN ECOLOGICAL COMMUNITY

The biological structure of an ecological community can be considered in terms of species diversity (including richness, evenness, abundance, dominance, and diversity at different spatial scales), the niches within the community, and trophic levels.

5.4.1 SPECIES DIVERSITY

At the core of the biological structure of a community are species composition (which taxa are present), species richness (how many of each species), species evenness (how uniformly individuals are distributed among taxa), abundance (the number of individuals of each species present), and dominance (which is defined differently according to the species concerned and the context).

5.4.1.1 Richness and evenness

Species richness is a count of the number of species present within a community (Figure 5.8). Timescale is important in recording species richness. For example, migratory birds are only present in a habitat for part of the year, hence recording species richness when the migrants are present or absent will give very different values.

Species evenness is the relative abundance of individuals among the species – the more equitable the distribution, the greater the evenness (Figure 5.9).

Several diversity indices have been developed to measure species richness and evenness. A key consideration when using these indices is their sensitivity to rare versus abundant species (Box 5.1). There are several statistical software packages (e.g. R or JASP) and online calculators which can be used to calculate richness, evenness, and diversity indices.

Figure 5.8 A highly diverse community of trees in Pai Canyon, Pai, Thailand (note the different leaf shapes indicating the diversity of plant species present).
Source: photograph by Gina Hazell

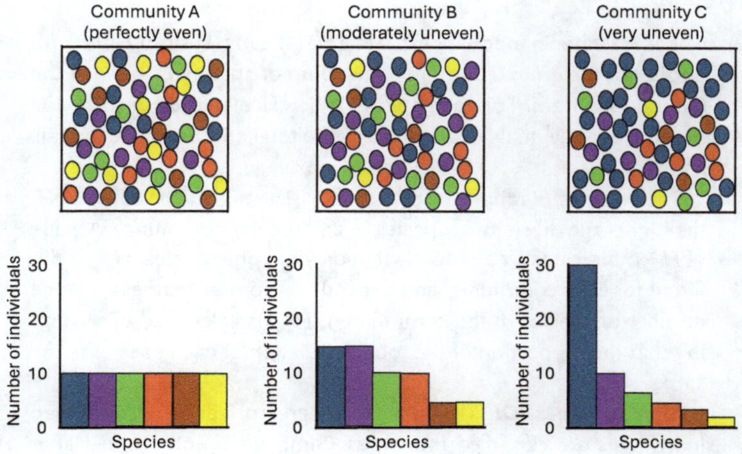

Figure 5.9 Three hypothetical communities with the same species richness ($N = 6$) but different evenness ranging from very even (community A) to very uneven (community C).

BOX 5.1 EXAMPLES OF DIVERSITY EVENNESS INDEXES

Richness index (R). A count of the number of different types of organisms within a community. This can be the number of species, but if it is not possible to identify all the species, some other methods of identification can be used (e.g. leaf shape). This index does not take into account the abundance of a species.

Shannon index (H). Also known as the Shannon diversity index and as the Shannon–Weiner index (Shannon 1948). It is named after Claude Shannon (1916–2001), an American who is credited as the 'father of information theory', and Warren Weaver (1894–1978), an American scientist, mathematician, and science administrator. It is calculated using the equation:

$$H = -\Sigma pi * ln(pi)$$

where H is the Shannon index, Σ (sigma) means 'sum', p_i is the proportion of the entire community made up of species i, and ln is the natural logarithm. For each species p_i is calculated by dividing the number of individuals in a species by the total number of individuals across allspecies.

The value of H ranges from 0 to $H_{max.}$ The higher the value of H, the higher the diversity of species in a particular community. A value of $H = 0$ indicates a community that only has one species. H_{max} is different for each community and depends on species richness (i.e. the number of species in the community). H_{max} is calculated by assuming that the proportion of the total population of each species is the same.

Simpson index (D). Measures the degree of concentration when individuals are classified into types (Simpson 1949). Named after Edward Hugh Simpson (1922–2019), a Northern Irish codebreaker, statistician, and civil servant. This index is calculated as:

$$D = \sum_{i=1}^{S} \left(\frac{n_i}{N} \right)^2$$

where n_i is the number of individuals in species i, N = total number of individuals of all species, and $n_i/N = p_i$ (proportion of individuals of species i), and S = species richness.

The value of Simpson's D ranges from 0 to 1, with 0 representing infinite diversity and 1 representing no diversity, so the larger the value of D, the lower the diversity. For this reason, Simpson's index is often presented as its complement $(1 - D)$. Simpson's dominance index is the inverse of the Simpson's index $(1/D)$.

Pielou's evenness index (J). Named after Evelyn Chrystalla Pielou (more commonly known as Chris Pielou) (1924–2016), a Canadian statistical ecologist who contributed significantly to the development of mathematical ecology and the mathematical modelling of natural systems.

Species evenness refers to how close in numbers each species is to the others. So, if there are 40 individuals of species A and 1000 individuals of species B, the community is not very even. But if there are 40 individuals of species A and 42 individuals of species B, the community is very even. The evenness of a community can be represented by Pielou's evenness index (Pielou 1966):

$$J = \frac{H}{Hmax}$$

where H is the Shannon index and H_{max} is the maximum value of H. The value of J ranges from 0 to 1. Higher values indicate higher levels of evenness. At maximum evenness, $J = 1$.

Pielou's evenness index (J) and Simpson index (D) can be used as measures of species dominance (the opposite of diversity) in a community. Low J indicates that 1 or few species dominate the community.

5.4.1.2 Abundance

Abundance considers the number of individuals of each species present. There are several methods for measuring abundance. Semi-quantitative abundance ratings involve a visual estimation of the abundance of each species in a specific area of a designated size. This is usually done in the field by using a quadrat of the appropriate size. Two semi-quantitative abundance ratings used are known as DAFOR and ACFOR (Table 5.1).

These scales are useful as qualitative estimates of the species abundance in a designated area, but they are not exact or objective measurements. A more quantitative method is to create a rank-abundance curve, also known as a Whittaker plot (named after

Table 5.1 The DAFOR and ACFOR semi-quantitative abundance rankings

DAFOR	*Species observed in an area is:*	*ACFOR*	*Species observed in an area is*
D	Dominant.	**A**	Abundant
A	Abundant.	**C**	Common
F	Frequent	**F**	Frequent
O	Occasional	**O**	Occasional
R	Rare	**R**	Rare

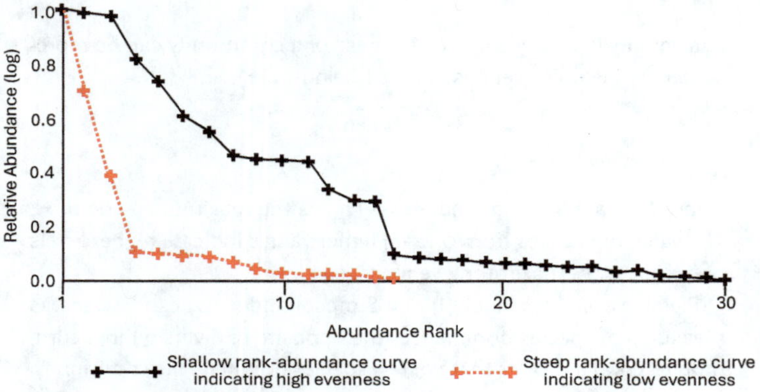

Figure 5.10 A rank-abundance (species) curve, also known as a Whittaker plot.

Robert Harding Whittaker (1920–1980, an American plant ecologist) (see Whittaker 1965) (Figure 5.10). This is done by counting the abundance of each species in a given area. The most abundant species is given rank 1, the second most abundant is 2 and so on. The species and their abundance are then plotted in decreasing order on a chart where the *x*-axis is the abundance rank. The *y*-axis is the relative abundance (e.g. the number of individuals relative to the abundance of all species and usually presented on a log scale).

A rank abundance curve depicts both species richness and species evenness. Species richness is the number of different species on the chart (in Figure 5.10, species richness (R) = 30 for the community represented by the black line, while R = 15 for the community represented by the red, dashed line). Species evenness is reflected by the gradient of the line that fits the graph. A steep gradient indicates low evenness (the high-ranking species have much higher abundances than the low-ranking species) and a shallow gradient indicates high evenness (the abundances of different species are similar). The curve also gives information about resource partitioning within a community as the plot can be viewed as a model of niche partitioning. That is to say, in a community where the organisms are of similar size, abundant species take a larger proportion of space and resources (i.e. a larger partition of the niche) than do less abundant species.

The max rank normalisation method is a technique used to transform species abundance data, to a common scale (Saeedghalati 2017). Each value in a dataset is divided by the maximum value within that same dataset. For example, if the maximum abundance of a species in a transect is 90 individuals, all abundance values for that species in that transect are divided by 90. The resulting normalised values range from 0 to 1, where a value of 1 represents the maximum abundance observed for that species, and 0 represents its complete absence. This standardisation helps to mitigate the effects of different sampling efforts or inherent differences in species abundance, making it easier to compare the relative importance or dominance of different species across various sites or samples. The simplicity of the max rank method makes it a useful tool for a variety of ecological analyses, including community comparisons and diversity studies.

5.4.1.3 Dominance

While it is common to refer to the dominance of certain species within a community, the concept of dominance is difficult to define. The term may be a reference to the most numerous species, the one with the highest biomass, the one that takes up most space, makes the largest contribution to energy flow and mineral cycling, or the one that controls (by some means or other) the other species in the community. To complicate matters further, in behavioural ecology the study of dominance and dominance hierarchies has long been of interest with the term dominance being used to indicate behavioural relationships between individuals and/or species.

To address the difficulty in identifying dominance, plant ecologists use the term relative dominance: described as a comparison of the number of individuals of one species compared to the number of individuals of all species in that population. Relative dominance is a ratio of either the biomass or basal area (in the case of trees, the cross-sectional area of a tree at a fixed height above ground – usually around 1.37 m – often referred to as breast height) of one species compared with the total for all species in the community. Plant ecologists also use the terms relative frequency and relative abundance. Relative frequency is an index based on the number of sample points or plots in which a species occurs relative to the total number of samples points or plots in the survey and is best used when there is a large difference in the size of individuals of the same species. Relative abundance is a measure of how common or rare a species is relative to other species in a defined location or community. Relative abundance and relative frequency can be combined to give an importance value index for each species. There are several indices that can be used to measure relative dominance, frequency, and abundance (Table 5.2).

Table 5.2 Dominance, density, frequency, and importance indexes used by plant ecologists

Index	*How calculated*
Dominance	$\frac{\text{basal area or aerial coverage, species A}}{\text{area sampled}}$
Relative dominance	$\frac{\text{basal area or aerial coverage, species A}}{\text{total basal area or coverage, all species}}$
Relative density	$\frac{\text{total number of individuals, species A}}{\text{total number of individuals, all species}}$
Frequency	$\frac{\text{intervals or points where species A occurs}}{\text{total number of sample plots or points}}$
Relative frequency	$\frac{\text{frequency value, species A}}{\text{total frequency value, all species}}$
Importance value	relative dominance + relative density + relative frequency
Simpson index of dominance:	$\frac{\Sigma n_i (n_i - 1)}{N(N-1)}$ where Σ means sum of, N is the total number of individuals of all species and n_i is the total number of individuals of species i
Importance value index (%)	relative abundance (%) + relative frequency (%) Note: As the importance value index is the sum of two percentages, it can range between 0% and 200%.

5.4.1.4 Diversity at different scales – alpha, beta, and gamma

Alpha (α) diversity (or species richness) refers to diversity within a single, relatively homogeneous ecological community or habitat. Beta (β) diversity measures the difference in species composition between two or more communities or habitats. Low beta diversity suggests that communities are similar in their species makeup while a high beta diversity means that two communities have very different species composition. Beta diversity helps in understanding how species composition varies across a landscape. Gamma (γ) diversity represents the overall diversity of species within a larger geographic region encompassing multiple habitats (e.g. a landscape, region, or even a biome) (Whittaker 1972) (Figure 5.11).

While they are useful for comparing different regions and identifying locations with high concentrations of native species that should be protected (i.e. biodiversity hotspots), alpha, beta, and gamma diversity do not completely account for genetic diversity, or ecosystem diversity and, hence, additional information may be needed when devising a conservation strategy.

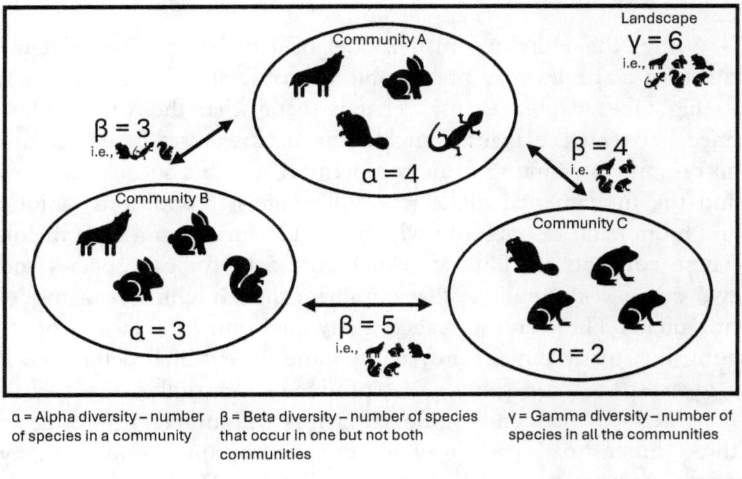

Figure 5.11 Alpha (α), beta (β), and gamma (γ) diversity – diversity at different spatial scales.

5.4.2 THE NICHE

The word niche, which is probably derived from the Middle French word *nicher* – meaning *to nest*, was first coined at the beginning of the twentieth century. Since then, the concept has been developed and described in different ways (Pocheville 2015) with definitions framed in terms of environments, habitats, roles, habitats and roles combined, trophic levels, n-dimensional spaces, resources, and requirements (Takola and Schielzeth 2022).

Three prominent approaches are those presented in the Grinnellian, Eltonian, and Hutchinsonian concepts of a niche (Table 5.3). The contemporary niche theory (also called 'classic niche theory') seeks to reconcile these three approaches. In this niche theory, the 'impact niche' is defined as the combination of effects that a given consumer has on both: a) the resources that it uses, and b) the other consumers in the ecosystem. The term 'requirement niche' is used to describe the range of environmental conditions (including both abiotic and biotic factors) where a species can successfully survive and reproduce (Letten et al. 2017). This theory does not, however, account for the invasion of non-native species into areas where they are introduced. There are, therefore, some issues around the concept of the niche that need to be resolved.

Within the Hutchinsonian model of a niche there is recognition that a species may not be able to utilise all the resources that it might be adapted to use. That is to say that there is a difference between the realised niche (the narrower set of conditions and resources resulting from competition which a species can use) and the fundamental niche (the full range of conditions (abiotic and biotic) and resources in which it could survive and reproduce). These concepts are illustrated in Figure 5.12 for two species and two of the *n*-dimensions that comprise the Hutchinsonian model of a niche. There are, in reality, many environmental factors operating (i.e. many dimensions) at the same time which determine if a given species can grow and reproduce successfully. It is easy to imagine how three environmental factors will interact to produce a three-dimensional space in which there are optimal conditions, say of pH, temperature, and nitrogen availability. When other factors are added in, such as phosphorous and magnesium concentrations,

Table 5.3 Grinnellian, Eltonian, Hutchinsonian, and contemporary concepts of the niche

Concepts of the niche	*Description*
Grinnellian (Grinnell 1917) (named after Joseph P. Grinnell (1877–1887) – an American field biologist and zoologist)	A niche is the sum of the habitat requirements and behaviours (the things a species 'needs') that allow a species to persist and produce offspring.
Eltonian (Elton 1927) (named after Charles Sutherland Elton (1900–1991) – a British zoologist and animal ecologist)	The place of an organism in the biotic environment (i.e. its relations to food and enemies). Conceptually, this introduces the idea that a species both responds to and effects the environment.
Hutchinsonian (Hutchinson 1957) (named after G. Evelyn Hutchinson (1903–1991) – a British ecologist)	A niche is an '*n*-dimensional hypervolume', where the dimensions are environmental conditions and resources (e.g. light, nutrients, temperature, moisture, wind, etc.) available to (and specifically used by) organisms and which define the requirements of an individual or a population to persist. Within this concept of a niche there are the ideas of a fundamental niche – the organism uses the full range of conditions (biotic and abiotic) and resources in which it could survive and reproduce – and of a realised niche – the narrower set of conditions and resources resulting from competition which a species can use.
Contemporary or classic niche theory	This framework centres around 'consumer-resource models' which splits a given ecosystem into resources (e.g. sunlight or available water in soil) and consumers (e.g. any living thing, including plants and animals), and attempts to define the scope of possible relationships that could exist between the two groups.

conductivity, and amount of light, it is not possible to illustrate what such an *n*-dimensional optimal space looks like.

George Gaylord Simpson (1902–1984) coined the term 'adaptive zone' to explain how a population could jump from one niche to another. That is to say that a species can jump into an 'adaptive

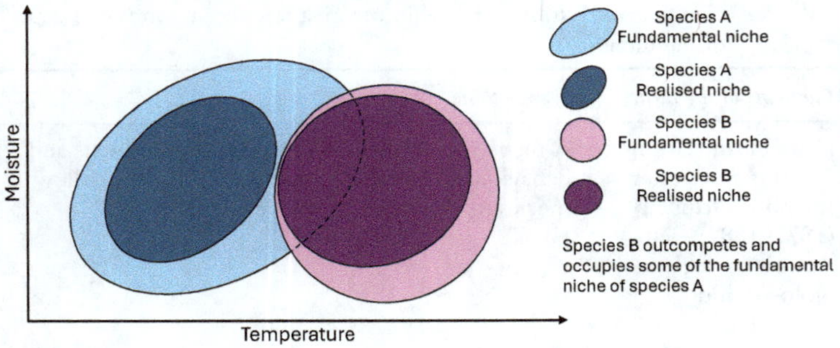

Figure 5.12 Fundamental and realised niches for two species and competition between these species.

zone', made available by some modification or change to which it is already adapted. This can explain how species can spread; they are already 'adapted' to the new ecological opportunity.

'Niche separation' describes the difference in the niche of two or more species. Where the niches are separate, the species have different requirements, referred to as 'niche partitioning' (e.g. one species of bird may feed on grain and another on insects). However, when two species use the same resource, for example an omnivorous bird (feeds on both plant material and insects) is present with an insectivorous bird there is 'niche overlap' and there is competition between the two species (in this case for the insects) (Figure 5.13).

An associated concept is that of 'niche partitioning', where species do not compete with each other (Figure 5.14). There is an assumption in Figures 5.13 and 5.14 that where niches overlap, competition not only can but does occur. This assumption is, however, not safe as the resource in the region of overlap may not be limiting, in which case there is no competition.

Organisms are often categorised as generalist or specialist based on their adaptability to ecological niches. This classification actually reflects a continuum from highly specialised to broadly generalist species. Generalist species are highly adaptable and have a wide niche (Chase and Leibold 2003). They can thrive in a wide variety of environments and utilise diverse food sources. Their flexibility allows them to adjust to changing conditions and disturbance (natural or caused by human). Examples of generalist species

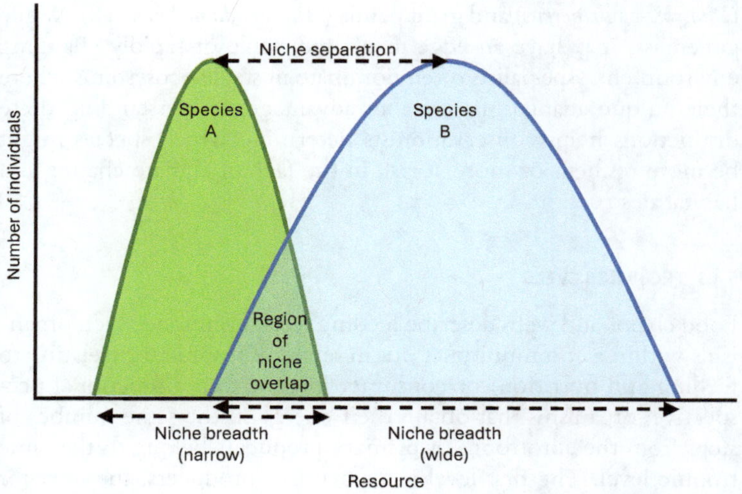

Figure 5.13 Niche width, separation, and overlap.

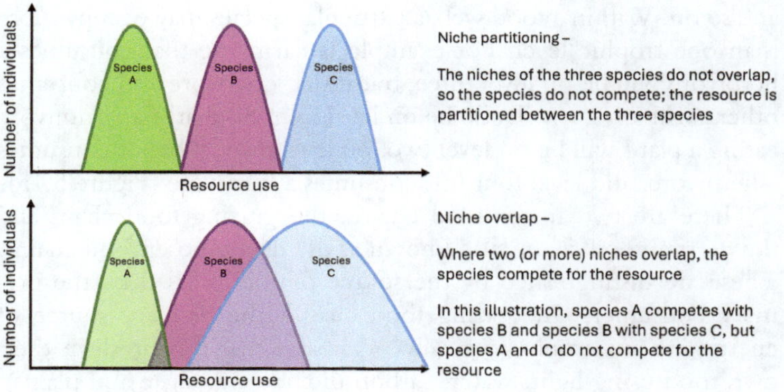

Figure 5.14 Niche partitioning and niche overlap for three species.

include raccoon (*Procyon lotor*) and red fox (*Vulpes vulpes*). In contrast, specialist species have a narrow ecological niche. They are adapted to specific environments or diets and can be highly efficient in those conditions. This makes them vulnerable to habitat changes or loss of a food source. Examples of specialist species include koala

(*Phascolarctos cinereus*) and giant panda (*Ailuropoda melanoleuca*). While generalists may have an edge in unpredictable or rapidly changing environments, specialists often dominate in stable ecosystems where their unique adaptations offer an advantage. Understanding these distinctions helps conservationists determine which species might be more resilient or more at risk in the face of climate change and habitat destruction.

5.4.3 TROPHIC LEVELS

Food chains and webs describe feeding relationships between organisms within a community set out in terms of trophic (i.e. relating to feeding and nutrition) or consumer levels. From a functional perspective, organisms that obtain their energy in the same number of steps from the autotrophs or primary producers belong to the same trophic level. The first level is for primary producers, the next for primary consumers (also referred to as herbivores), the third level is for secondary consumers (also known as carnivores), the next level is for tertiary consumers (i.e. carnivores that eat other carnivores) and so on. Within a food web, a particular species may occupy more than one trophic level. For example, a carnivore that consumes a herbivore will be on level three, but if that carnivore also consumes other carnivores, it will also be on level four. Similarly, an omnivore eating a plant will be on level two, on level three when it consumes a herbivore, and level four if it consumes a carnivore (Figure 5.15).

There are two major food chains: the grazing food chain and the detrital food chain (also known as the decomposer food chain). These are distinguished by the source of energy used by the primary consumers. In grazing food chains, the primary source of energy are autotrophs (typically organisms that can produce their own food using light, water, carbon dioxide e.g. algae and plants). In contrast, the primary source of energy in detrital food chains is dead matter and detritus of any origin. The two main components of the detrital food chain are decomposers and detritivores. These two classes of organisms obtain their energy through different approaches. Decomposers, which include bacteria and fungi, are organisms that break down, organic material such as the remains of dead organisms through biochemical reactions which are external to their bodies. For example, they secrete enzymes which break

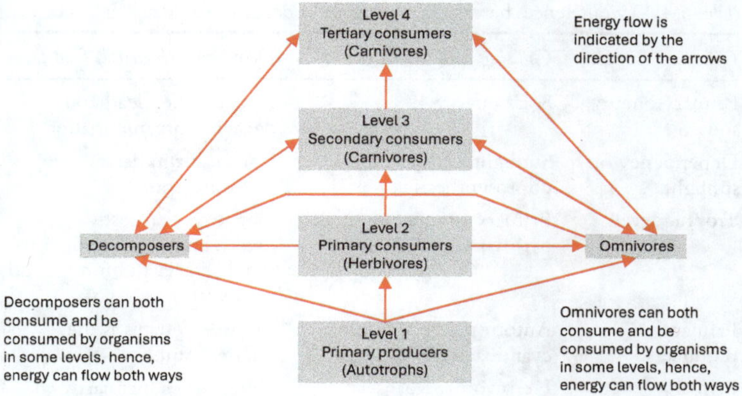

Figure 5.15 Schematic, simplified food chain illustrating the flow of energy through a community.

down dead plant and animal material and then absorb the nutrients released. Detritivores, in contrast, must digest organic material within their bodies to break it down and gain nutrients. Examples of detritivores include invertebrates such as sea stars (Asteroidea), fiddler crabs (Ocypodidae), slugs and snails (Gastropoda), earthworms (Annelida), and woodlice (Oniscidea). There are some important differences between the grazing and detrital food chains (Table 5.4).

There are interactions between the grazing and detrital food chains (Figure 5.16). One such interaction is through predation: carnivores in both food chains predate herbivores in the grazing food chain and decomposers and detritivores in the detrital food chain.

There is a third food chain, the parasitic food chain. Illustrating this chain is complicated by the complex life cycle of many parasites involving different stages in their life cycle and primary and secondary hosts. Parasitism constitutes one of the most ubiquitous and diverse ecological interactions, occurring in virtually every habitat on earth and involving organisms from viruses and bacteria to fungi, protozoa, helminths, and arthropods. Hence, there are multiple parasitic food chains within each component of the grazing and detrital food chains shown in Figure 5.16. Some parasites alter host behaviour, physiology, and population dynamics, thereby,

Table 5.4 Key differences between a grazing and a decomposer (detrital) food chain

Criteria	*Grazing food chain*	*Decomposer (detrital) food chain*
Primary energy source	Solar energy	Detritus (i.e. dead and decaying organic matter)
Dependency on sunlight	Sunlight is required for photosynthesis	Can occur in dark environments
Trophic levels	Primary producers, herbivores, and carnivores	Detritivores (mostly invertebrates), decomposers (i.e. bacteria and fungi), and carnivores
Primary producers	Autotrophs (plants, algae, cyanobacteria)	None – detritus is the energy source
Primary consumers	Herbivores	Detritivores (primarily invertebrates)
Direction of energy flow	Unidirectional – from primary producer, through herbivores to one or more tropic levels of carnivores	Bidirectional – energy contained within plant and animal waste and dead bodies passes back to decomposers and detritivores
Efficiency in the transfer of energy	Approximately 10 per cent of energy is transferred between each trophic level	Up to 30 per cent of energy is transferred between each trophic level
Contribution to nutrient cycling	Limited	Very significant

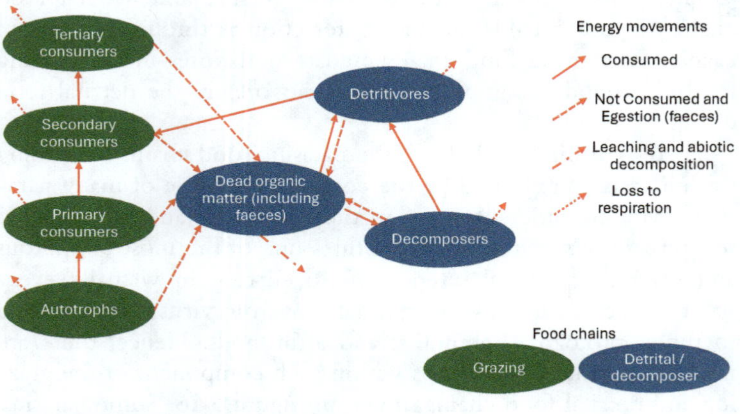

Figure 5.16 Interactions between grazing and detrital food chains illustrating the connections between the two food chains.

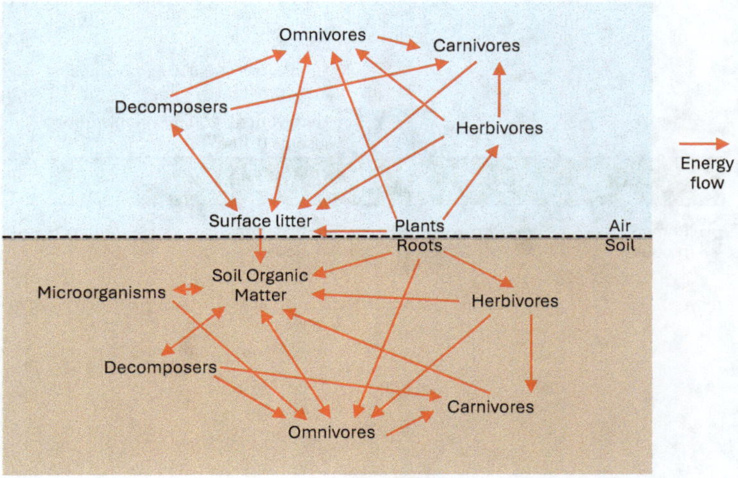

Figure 5.17 Conceptual grassland food web.
Source: redrawn from Buzhdygan et al. (2020)

shaping community structure, driving coevolutionary processes, and contributing to ecosystem resilience and stability (which will be discussed in Chapter 6).

Food webs, which integrate several food chains, are complex and include both above and below ground components (Figure 5.17).

The structure of a food web can be visualised as a pyramid with trophic levels stacked one on the other. Such an approach was first described by Charles Elton (biographic details in Table 5.3) in 1927. Ecological pyramids show the relative amounts of various parameters (such as number of organisms, biomass, and energy) in each trophic level. The pattern produced varies according to the parameter used, the ecosystem, and in some cases, the season.

Using the number of individuals in each trophic level produces pyramids that can appear as either upright or inverted depending on the community and season. During the summer a typical grassland pyramid has a base of many plants, and the number of organisms decreases at each trophic level (Figure 5.18). In comparison, in a temperate forest in the summer, the base of the pyramid consists of the trees. Feeding on each tree is a large number of primary consumers (e.g. aphids). In temperate climates there are many fewer

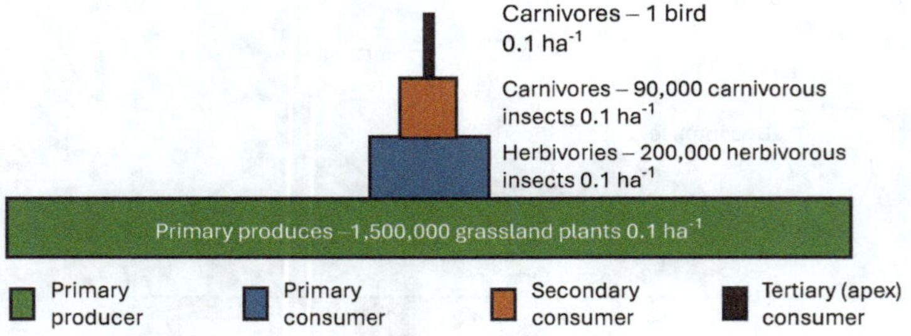

Figure 5.18 Typical pyramid of organism numbers in a hypothetical temperate grassland in summer.

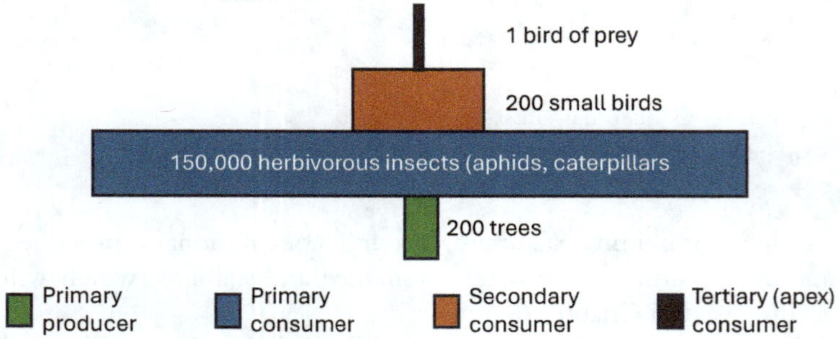

Figure 5.19 Inverted pyramid of organism numbers in a hypothetical woodland illustrating how many individuals may be dependent on one individual of a different species.

insects in winter and hence, a pyramid of numbers at that time will be very different in shape to that in the summer (Figure 5.19). This may not give a true reflection of the structure of the trophic levels.

Another parameter that can be used to construct an ecological pyramid is biomass. Biomass is a measure of the amount of energy converted into living tissue at the different trophic levels. Not all the energy in one trophic level is passed on and incorporated into the bodies of the organisms in the next trophic level (this will be discussed further in Chapter 6). As with a pyramid of numbers, the biomass pyramid may be upright (Figure 5.20) or inverted (Figure 5.21). In the example in Figure 5.21, phytoplankton make

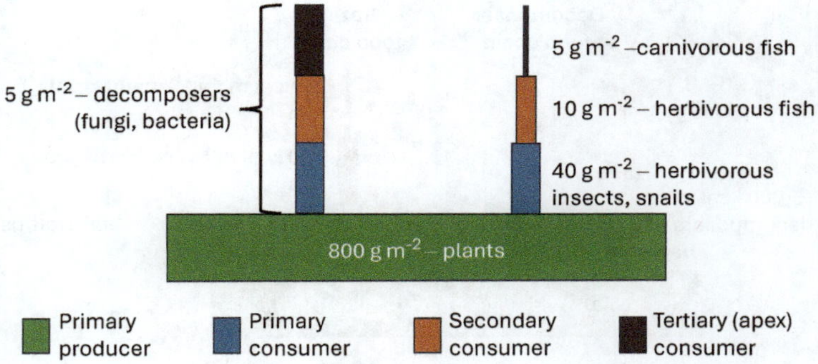

Figure 5.20 Pyramid of biomass for an aquatic food chain based on plants.

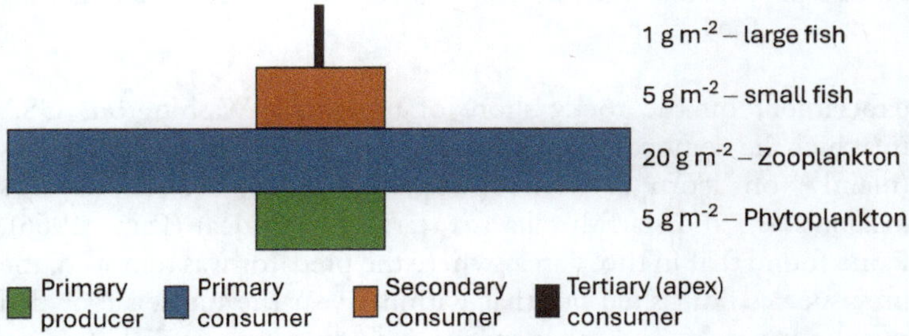

Figure 5.21 Inverted pyramid of biomass for an aquatic food chain based on plankton.

up less biomass than the primary consumers (i.e. the zooplankton) due to the high turnover rate of the phytoplankton (i.e. while there is a high reproduction rate each individual phytoplankton does not live for very long before it is consumed or dies).

The final way of visualising a food web is with a pyramid of energy. In contrast with pyramids of numbers and of biomass, pyramids of energy are always upright as there is also less energy stored in each higher trophic level (this point will be discussed in more detail in Chapter 6) (Figure 5.22).

Until the 1960s it was generally thought that all species on a given ecosystem were equally important. In that decade Robert T. Paine (1933–2016, an American zoologist) conducted an invasive

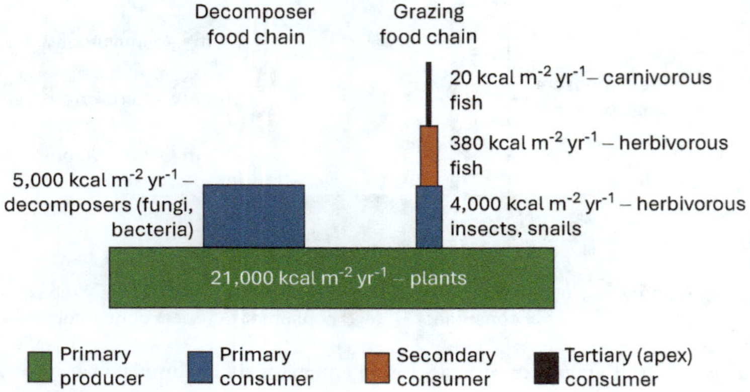

Figure 5.22 Pyramid of energy in a hypothetical decomposer and a grazing food chain.

experiment off the rocky shore of northwest Washington, USA in which he removed the predatory starfish (*Pisaster*) which preys (mainly) on acorn barnacles (*Balanus glandula*), mussels (*Mytilus californianus*), and starfish (aka sea stars) (Asteroidea) (Paine 1966). Paine found that in those areas where the predator was removed, the prey species flourished but that within a year the mussels crowded out the barnacles and other species including limpets and algae on which many other species depended resulting in a halving of the number of species in the ecosystem. Paine recognised that the predatory species was controlling the structure of the community and in 1969 he published a paper in which he called these highly influential predators 'keystone species' (Paine 1969). This term references the central stone in an arch (i.e. the keystone) which maintains the arch's stability and keeps it from falling.

As well as the concept of keystone species Paine also developed the concept of a trophic cascade to describe how the ripples of the change at one trophic level spread through other levels. The cascade may spread from a top predator down or permeate up through a food web as a result of changes in the availability of nutrients and other abiotic factors. For example, the introduction of a carnivore predator will reduce the population size of its prey, and, hence, reduce grazing pressure which allows for more plant growth.

Developments from Paine's work led to the concepts of foundation species and umbrella species. In 1972 Dayton coined the term foundation species to describe an organism that plays a singular, defining role in structuring an ecological community, often by creating or maintaining the physical habitat itself. Unlike keystone species, which exert influence disproportionate to their abundance, foundation species are typically the most abundant organisms in a system and possess the highest biomass. By stabilising the environment and reducing physical stress, foundation species facilitate the presence of a diverse array of other organisms. Classic examples include corals (e.g. *Acropora* spp.) in reef ecosystems, which build the calcium carbonate structures that house thousands of marine species, and eastern hemlocks (*Tsuga canadensis*) in North American forests, which regulate soil chemistry and canopy shading to create a unique microclimate (Ellison et al. 2005). The loss of a foundation species often precipitates a fundamental shift in the ecosystem's state, leading to a collapse of the associated community structure.

Wilcox (1984) (American conservation ecologist) formalised the concept of an umbrella species as one that is selected for conservation planning due to the protection it indirectly confers to many other naturally co-occurring species. This concept relies on the premise that these species – often large-bodied mammals or birds – have such extensive range requirements or specific habitat needs that protecting sufficient land for them automatically protects the habitats of numerous smaller, less demanding species. In this way, umbrella species act as 'surrogates' for the broader biodiversity of a region, allowing conservationists to maximise results by focusing resources on a single target species. A prominent example of an umbrella species is the tiger (*Panthera tigris*). Maintaining large tracts of forest in Asia for a viable tiger population preserves the ecological integrity of entire watersheds and the species within them.

5.5 ECOLOGICAL COMMUNITY DYNAMICS

Communities are dynamic and tend to change along a predictable trajectory. This trajectory is termed succession. In the aquatic environment of a pond, the process of succession is from open water to dry ground and is typically driven by the inflow of nutrients, organic matter, and pollutants as well as changes in the physical

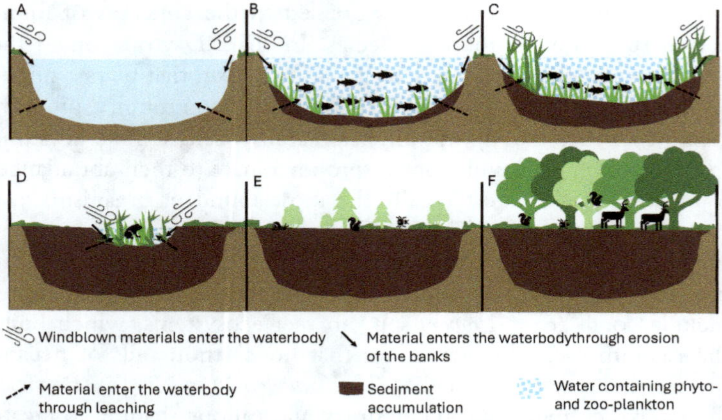

Figure 5.23 Succession of a pond or lake from open water to dry land.

environment including temperature and sunlight, and the species present and their interactions (Figure 5.23).

In terrestrial environments, a distinction is made between primary and secondary succession. Primary succession takes place on pristine substrates – substrates that have not been previously altered by physical, chemical, or biological (includes physical biological and chemical biological) weathering (Table 5.5). Such substrates include volcanic larva, sand dunes, rock exposed by a landslip or by the erosion of rocky outcrops and cliffs to reveal unweathered rock, newly exposed glacial till, and on construction materials including cement, concrete, and brick (Figure 5.24). The general trend in primary succession is from bare rock, through various stages before arriving at the climax community that is relevant for the geographical area. Figure 5.25 illustrates succession ending in forest. Where forest is not possible due to the climate (see Chapter 2), succession may end at grassland, shrubland, or another habitat.

Where substrates have undergone weathering of any type, the process of succession that takes place is termed secondary succession. Typically, secondary succession occurs when vegetation is cleared from an area (Figure 5.26). Natural and human disturbance resulting in the loss of vegetation can occur at any stage of succession and succession can be reset at any earlier stage.

Table 5.5 The impact of types of weathering on the environment

Type of weathering	*Description*
Physical	Wind or water blown particles acting on a surface or of rocks, stones. Pebble being moved by water currents. Freeze/thaw cycles in which water penetrates fissures in the rock, freezes and expands to open the cracks still further or break off a piece of rock.
Chemical	Chemicals in the environment, for example acid rain wearing away the surface of a rock.
Biological	
Physical biological	Occurs when a force or pressure exerted by an organism breaks rocks apart or degrades the minerals in them. Examples of organisms involved in this process include: plant roots growing on and into rocks; burrowing animals like shrews (Soricidae), moles (Talpidae), earthworms (Annelida), and ants (Formicidae); and birds creating holes and eroding the upper surface of the soil as they forage for seeds and earthworms.
Chemical biological (also called organic weathering)	Occurs when living organisms excrete organic compounds containing molecules that acidify and corrode rock minerals. Examples include limpet (Patellogastropoda), piddock (a bivalve mollusc; *Pholas dactylus* and *Barnea candida*), the fungal component of lichen, plant roots, and fungi that produce siderophores (a type of chelating agent which can absorb various minerals and nutrients from the soil).

(A) (B)

Figure 5.24 Examples of new material which has not been weathered and, hence, a site for primary succession: volcanic rock on (A) Mount Teide, Tenerife, Canary Isles and (B) Mount Etna, Sicily, Italy.

Sources: (A) photograph by Carole Conroy; (B) photograph by Sue Melville

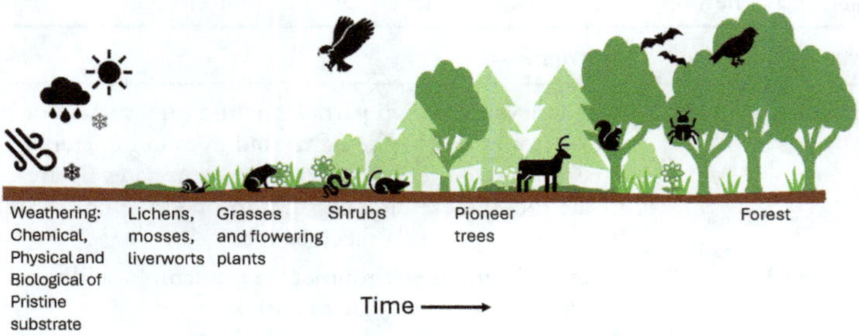

Figure 5.25 Primary succession – from unweathered rock to a forest.

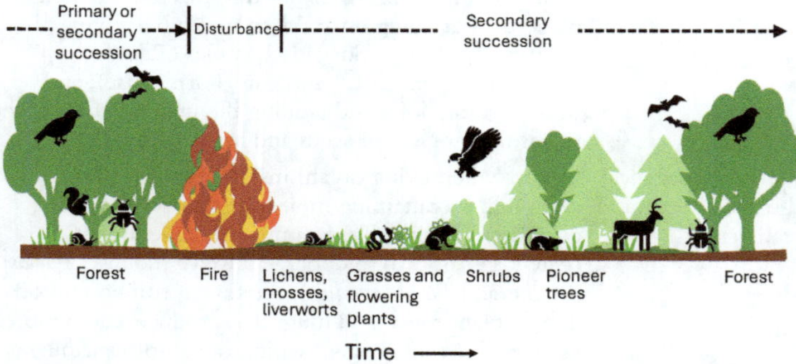

Figure 5.26 Secondary succession – from soil that had previously been weather during primary or secondary succession to a forest.

A fire (Figure 5.27), landslide, earthquake, high wind event, and flood are natural events that can lead to the removal of vegetation, taking the community back to an earlier successional stage. The same is true of habitat management techniques that involve the cutting, pruning, or removal of vegetation. For example, mowing or grazing a semi-natural grassland prevents the grassland moving to a shrub stage and eventually to a forest.

Some plants are adapted to require a disturbance before they complete their life cycle. For example, pyrophily is a strategy in which the plants require fire to complete their life cycle. Such plants are found in fire-prone ecosystems (e.g. Mediterranean scrublands or Australian heathlands). In Mediterranean regions, the seeds of

Figure 5.27 Secondary succession in (A) a forest in Scotland where a fire occurred two weeks prior to this photograph, and (B) two months after a fire had swept through an area of bilberry (*Vaccinium myrtillus*) on a moor in the Peak District, England.

Sources: (A) photograph by Sue Melville; (B) photograph by Barbara Firth

Cistus shrubs break dormancy only after being exposed to heat or smoke and in Australia, species of *Banksia* trees have woody cones that open only after exposure to intense heat ensuring that when the cones open, the seeds are releases into nutrient-rich ash beds.

In both primary and secondary succession the trajectory of change is influenced by factors including the initial site conditions, the type of disturbance, interactions of the species present, and by more random factors such as the arrival of seeds or colonists and weather conditions at the time of disturbance. Hence, while some aspects of succession are broadly predictable, there are aspects that are unpredictable (e.g. the actual species present, and species richness and abundance). It is also not generally possible to predict at what point natural disturbances will take place along a successional trajectory.

Irrespective of whether succession is primary or secondary, there are three phases recognised in succession: colonisation, successionist, and climax. Organisms that colonise newly available areas are referred to as early successional species (Table 5.6). Early successional stage plants are described as pioneer spices. They are generally fast-growing, well-dispersed, opportunist, fugitive species which typically have a *r*-selected life-history species (see Chapter 4). Early successional plants stabilise soil, alter the levels of soil organic matter, nutrient, and moisture, and provide shelter from water currents or from winds and shade from the sun. Species richness nearly always increases during early succession as new species arrive. This stage sets the foundation for the successionist phase during which late successional species become more dominant (Table 5.6). Pioneer species are replaced by more competitive (*K*-selected) species. Species richness, overall plant biomass, plant lifespans, the importance of decomposer organisms, and overall stability all increase. At the same time the rate at which soil nutrients are consumed, biogeochemical cycling, and net primary productivity all decrease as a community approaches the climax phase (Emery 2010). This climax phase is relatively stable and will persist until a disturbance occurs. Species richness tends to be lower in this stage than at earlier

Table 5.6 Physiological characteristics of early and late successional plants

Attribute	*Early-stage successional plants*	*Late-stage successional plants*
Seed dispersal in time	Well dispersed	Poorly dispersed
Seed germination enhanced by:		
Light	Yes	No
Fluctuating temperature	Yes	No
High nitrate	Yes	No
Seed germination inhibited by:		
Far red light	Yes	No
High CO_2 concentrations	Yes	No
Light saturation intensity	High	Low
Light compensation point	High	Low
Efficiency at low light	Low	High
Photosynthetic rates	High	Low
Respiration rates	High	Low
Transpiration rates	High	Low
Stomatal and mesophyll resistances	Low	High
Resistance to water transport	Low	High
Recovery from resource limitations	Fast	Slow
Resource acquisition rates	Fast	Slow

Source: after Bazzaz (1979)

stages of succession as competition eliminates opportunistic species and leads to dominance of the community by species that can outcompete others.

Zonation is common in both aquatic and terrestrial environments and can be associated with variations in climate factors associated with altitude (the community at the bottom of a mountain is not the same as that at the top), changes in salinity (the estuarine community varies as the salinity changes from high to low along an estuary), moisture level (the community changes moving inland from the edge of a lake), and exposure (the community on a rocky shore changes with increased distance up the shore). The word sere (from series) is sometimes used to describe the collection of different stages of zonation, for example, the zonation on a salt marsh may be referred to as a halosere (i.e. ecological succession in saline aquatic environments). The word sere is also used to describe changes over time. Figures 5.25 and 5.26 also illustrate a lithosere (i.e. plant succession that begins on a newly exposed rock surface). Figure 5.28 illustrates a hydrosere (i.e. ecological succession that occurs in an aquatic environment).

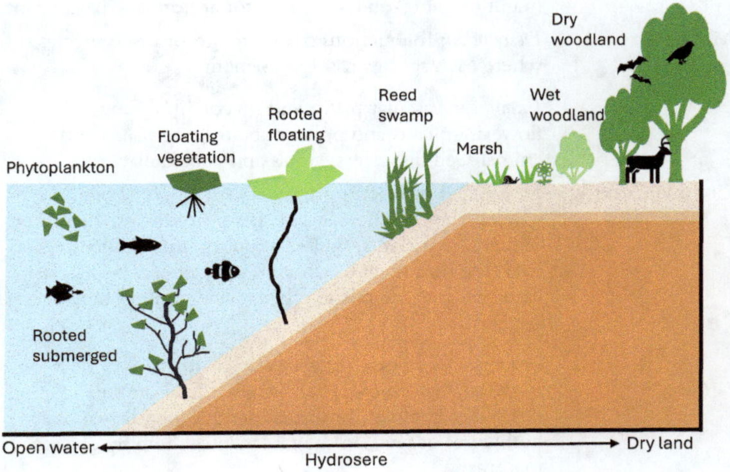

Figure 5.28 A hydrosere – from open water to dry woodland.

5.6 ECOLOGICAL COMMUNITY INTERACTIONS

5.6.1 TYPES OF INTERACTIONS BETWEEN SPECIES

Symbiosis, is a term covering close, long-term biological interaction between two individuals of different species, termed symbionts. The different types of symbiotic interactions can be described according to the effects on the species involved. These effects may be neutral (0), positive (+), or negative (–) (Table 5.7).

Symbiosis can be obligatory (i.e. one, or both symbionts depend on each other for survival), or facultative (optional), where the symbionts can also thrive independently. Symbiosis is also classified

Table 5.7 Characteristics of interactions between species – positive (+), negative (–), and neutral (0) responses in the population growth of two interacting species A and B

Type of interaction	*Description and examples*
Neutral 0 / 0	Description: Interactions have no effect on the population of either species. Examples: Forest monkeys often associate in 'polyspecific associations' (i.e. in groups comprising two or more species). In such groupings some species' relationships might be more-or-less neutral for at least one participant.
Mutualism + / +	*Description*: Interactions between two or more species where each species has a net benefit. *Examples*: Vascular plants and mycorrhizal fungi, flowering plants and pollinators, fruiting plants and animal seed dispersers, corals and zooantellae.
Commensalism + / 0	*Description*: Interaction in which members of one species (the commensal) benefits from the association by obtaining nutrients, shelter, support, and/or locomotion from the host species, which is substantially unaffected gains while those of the other species neither benefit nor are harmed. *Examples*: Remoras (Echeneidae) that ride attached to sharks (Selachimorpha), fish that feed on their hosts' faecal matter, birds that perch on bodies of large herbivores or feed on the insects turned up by grazing mammals.

(*Continued*)

Table 5.7 (Continued)

Type of interaction	*Description and examples*
Amensalism – / 0	*Description*: Interactions in which one species is inhibited or killed and the other is unaffected. There are two types of interaction: • Competition –in which the same resources are used by different species. • Antibiosis (aka allelopathy) – substances produced and released by one species, including acids and bases, simple organic compounds reduce competition for nutrients, light, and space, prevent or restrict growth in another species. Examples: Broomsedge (Andropogon virginicus) produces chemicals that inhibit other shrubs from growing. Bracken fern (Pteridium aquilinum) produces phytotoxins that accumulate in the upper surface of soils and kill germinating seeds of many plants.
Parasitism + / –	*Description*: Interactions in which one organism, the parasite, lives on or inside another organism, the host, causing it some harm, and is adapted structurally to this way of life. *Examples*: Viruses; bacteria; single-celled protozoans such as the agents of malaria (*Plasmodium*), sleeping sickness (*Trypanosoma brucei*), and amoebic dysentery (*Entamoeba histolytica*); fungi (e.g. honey fungus (*Armillaria*), ringworm (caused by about 40 fungi)); plants (e.g. European mistletoe (*Viscum album*), dodder (Cuscuta), and broomrapes (Orobanche)); and animals (e.g. helminths (including roundworms, tapeworms, and trematodes), lice (Phthiraptera), and vampire bats (Desmodontinae)).
Predation + / –	*Description*: One organism, the predator, kills and eats another organism, the prey. *Examples*: Carnivorous plants, some fungi (which produce specialised structures covered with sticky substances) have specialised cells that form constricting rings or produce chemicals that attract their prey which they then trap), and animals.

by physical attachment. Conjunctive symbiosis is when the symbionts form a single body (e.g. lichens (a symbiosis between algae or cyanobacteria and fungi) and coral reefs (a symbiosis between corals – colonial marine invertebrates – and algae), while all other arrangements are called disjunctive symbiosis. Within disjunctive symbiosis there is ectosymbiosis in which one organism lives on the surface of another organism and endosymbiosis where one symbiont lives inside the tissues of another organism (e.g. coral) (Paracer and Ahmadjian 2000; Moran 2006).

5.6.2 HERBIVORE–PLANT INTERACTIONS

Herbivore-plant interactions can be to the benefit of the plant, the herbivore, or both. In general, herbivores feed on vegetation – eating seeds, fruits, leaves, bark, stems, sap, and/or roots (Figure 5.29). Ecologically, this is a form of predation (see Table 5.7). Consumption by herbivores of leaves, bark, stems, sap and/or roots is destructive to a greater or lesser extent. Large-scale removal of plant biomass adversely affects a plant's vigour and ability to compete with other plants. If grazing continues, the plant may eventually die. However, light to moderate grazing can be beneficial to the plant as it may have a stimulating effect resulting in a flush of new growth. This is especially the case with grasses (Poacae). The growth point of a grass is at the base of the leaf and removal of a leaf stimulates new growth and tillering (i.e. the production of more leaves). Plants can derive other benefits from herbivores. Herbivorous birds and mammals transport seeds from one location to another on and in their

(A) (B)

Figure 5.29 Predation on plants by herbivores browsing and grazing. (A) A Masia giraffe (*Giraffa tippelskirchi*) browsing in the Masai Mara National Park, Kenya. (B) A llama (*Lama glama*) grazing on the Lars Trail in the Peruvian Andes.

Source: photographs by Rob Young

bodies. Seeds and fruit that pass through an animal's digestive system are exposed to acids, which weaken seed or nut casings. This is an example of scarification, a process that results in the weakening or opening of seed coats or nuts. Scarification can be by chemicals, cold temperatures, the passage of time, and mechanical abrasion. In all cases seed germination is aided. Seeds and nuts that pass through an animal's digestive system may also benefit from being deposited along with nitrogen rich guano or faeces, providing nutrients for germination and early growth.

Plants have a range of defence mechanisms – structural, predator satiation, mimicry, and chemicals – against herbivores (Table 5.8).

Plants are well-connected and send messages to their peers via their roots (Elhakeem et al. 2018) and an underground network of fungal mycorrhiza (first identified in 1997 by Suzanne Simard (1960–), a Canadian forest scientist and conservationist) and now commonly known as the 'wood wide web'). This web enables water

Table 5.8 Plant defence mechanisms against herbivory

Defence mechanism	*Description*
Structural defences	Includes trapping, impaling, fencing out, or discourage browsing using tough leaves, thorns, spines, hard seed coat or nuts, and epidermal hairs (which may sting).
Predator satiation	Occurs when an abundance of food (e.g. seeds) is produced over a short time (i.e. masting) in such an amount that the predators cannot consume it all. Four strategies for predator satiation have been identified (Janzen 1971): 1. Distribute seeds so that most fall near the parent tree while some fall at some distance. Many seed predators tend to congregate around the parent tree, where the density of seeds is highest, the more widely scattered seeds are not found and eaten. 2. Shorten the time of seed availability. In this approach, seed predators cannot consume all the seeds within the time available. 3. Synchronisation within a species so that seeds are produced periodically rather than annually. This strategy relies on the number of seed predators falling in years when there are few or no seeds. 4. Synchronise fruiting or seed drop between species that share the same seed predators.

(*Continued*)

Table 5.8 (Continued)

Defence mechanism	*Description*
Mimicry	One organism mimics the appearance of another. For example, (1) some passionflower (*Passiflora* spp.) leaf shapes mimic those of plants that the herbivorous passionflower butterfly (*Heliconius*) finds inedible (Gilbert 1975). (2) *Passiflora cyanea* and *P. auriculata* have glandular outgrowths on the stipules (structures near the leaf base) that mimic the size, shape, and colour of *Heliconius* eggs. *Heliconius* females avoid laying eggs on shoots that already carry eggs; hence, egg laying is deterred (Williams and Gilbert 1981).
Chemical defences	Production of secondary metabolic products (i.e. chemicals produced in the plant that are not directly related to plant metabolism), including alkaloids, terpenes, phenolics, resins, steroids, cyanogenic, mustard oil glycosides, and tannins. These compounds may be released when the cells in which they are stored are broken, (e.g. when a herbivore bites into a leaf), or they may be stored in epithelial cells and secreted by epidermal glands. These compounds may act as warning orders, repellents, inhibitors, attractants, or, in some cases, direct poisons. Sometimes a substance in, or secreted by, one plant serves to give protection to others (e.g. acacia trees (*Acacia* spp.) responding to giraffe (*Giraffa* spp.) predation).

and nutrients to be redistributed from resource-rich to resource-poor areas, and the transmission of chemical warning signals against pathogens or herbivores from one plant to another. There is even preferential carbon transfer to offspring or close relatives. Some chemical defences are activated only when a plant senses that it or a neighbour is being attacked by insects (Angrish 2022). It has also been found that plants responsive behaviour, such as defence responses and gene regulation, can be altered by their fungal network (Gorzelak et al. 2015). It is believed that the network of fungi hyphae connects 80 per cent of plants (Castro-Delgado et al. 2020). Mycorrhiza fungi enhance plant diversity and productivity, stabilise soil structure, accelerate decomposition, and drive carbon sequestration in terrestrial ecosystems (Simard et al. 1997).

Some grazers have developed mechanisms to counter the effect of the defensive chemicals produced by plants. The main mechanism employed is detoxification, whereby the secondary metabolic products produced by plants are acted on by the consumer's mixed function oxidase system. This system involves oxidation, reduction, and hydrolysis and converts fat soluble chemicals (many of the defensive chemicals produced by plants are fat soluble) into water soluble ones that can then be excreted. The mixed function oxidase system is a non-specific, general detoxifying system, which is activated by a wide range of toxic compounds. This means that new toxic substances can be detoxified with little genetic change necessary (Brattsten et al. 1977).

Another strategy used by some insects feeding on leaves that contain toxic latex and resins is to stop the flow of that sap to the areas they are feeding on. This is achieved by cutting leaf veins to sever the secretory canals, blocking the flow of latex to the intended feeding site, or if feeding on plants from the bindweed or morning glory family (Convolvulaceae), in which the canals are confined to the major leaf veins, avoiding these veins when eating. The larvae of monarch butterfly (*Danaus plexippus*) feed on milkweeds (*Asclepias* spp.). Milkweed latex contains cardenolides – which are cardiac glycosides and affect the force with which heart muscle contracts) – as well as sticky resins that can immobilise or poison herbivores. Monarch butterfly larvae have behavioural and physiological strategies to cope with this toxic latex. Behaviourally, monarch larvae selectively feed on young leaves, where latex pressure is lower, and employ 'trenching' or 'vein-cutting' behaviour: they chew through the leaf's main veins before feeding, reducing latex flow to the targeted area. This is a behaviour that is also observed in other insects that feed on latex producing plants (Dussourd and Eisner 1987; Dussourd and Denno 1991). Physiologically, monarch butterflies are adapted to tolerate and sequester cardenolides. They store the toxins in their bodies which then act as a deterrent against predators (Zalucki and Malcolm 1999).

5.6.3 ANIMAL PREDATOR–PREY INTERACTIONS

While herbivores face challenges in obtaining good quality food (in terms of the nutrients contained in the food) and, in appropriate quantities, carnivores face only the challenge of catching enough prey as the quality of their food is generally good (Figure 5.30). Prey animals have strategies to avoid predation, and predators have strategies to capture prey (Table 5.9).

(A) (B) (C)

Figure 5.30 Predation on animals. (A) Nile crocodile (*Crocodylus niloticus*) in Kenya. (B) Female African lions (*Panthera leo*) with their prey in the Masai Mara National Park, Kenya. (C) Female Eurasian sparrowhawk (*Accipiter nisus*) with its prey, a common wood pigeon (*Columba palumbus*), on the author's front lawn in Cheshire, England.

Sources: (A) photograph by Sue Melville; (B) photograph by Rob Young; (C) photograph by P. James

Despite these defences, predators are very successful at catching prey. Predation is the leading cause of death in small animals and juveniles. Juvenile mammals and reptiles are twice as likely as adults to be eaten by predators. More than 95 per cent of documented deaths of juvenile reptiles were due to predation. Birds seem to be the exception to this rule: juveniles are no more likely than adults to be eaten, and overall, causes of death are similar across life stages (Hill et al. 2019).

Carnivorous predators have three basic methods of hunting: ambushing, stalking, and pursuit. The tactics can be placed on a continuum of time maximising to energy maximising. Pursuit hunting, for example, requires a lot of energy to be used in a short time. An example of a pursuit hunter is a lion (*Panthera leo*) (see Figure 5.31).

Table 5.9 Animal defence mechanisms against predators

Defence mechanism	*Description*
Predator satiation	Some animals produce more eggs or young than their predators can consume in a short period of time. For example, corals (colonial marine invertebrates within the class Anthozoa), many species of fish such as salmon (Salmonidae), and ungulate mammals such as caribou (*Rongifer tarandus*) and wildebeest (*Connochaetes taurinus*). Reproductive synchrony is taken a step further in the 13-year and 17-year appearances of the periodic cicadas (*Magicicada* spp.).
Armour and weapons	Includes the protective armour of clams, armadillos, turtles, and numerous insects, and other protective structures such as the quills (which are modified hairs) of porcupines (Hystricognathi), hedgehogs (Erinaceinae), echidnas (also known as spiny anteaters – Tachyglossidae) and some fish.
Behaviour	Prey animals can hide, run away, or fight predators. Some animals employ a distraction display, for example, birds with precocious young (young are well-developed when they hatch and leave the nest within a day or two of hatching) use distractive behaviours at the time that the eggs hatch, while altricial birds (young are born naked and helpless), use the behaviour at the time the young fledge.
Camouflage	Cryptic colouration (including colour pattern), shape, posture, movements, and behaviours or a combination of these. is common among invertebrates, fish, reptile, many ground-nesting birds, and mammals.
Warning colouration and Mimicry	Warning colouration is used by some animals as a signal that they are toxic or have a chemical defence that other animals find repulsive. There are different types of mimicry: 1. Batesian mimicry (named after H.W. Bates (1825–1892), an English explorer-naturalist) – the resemblance of an edible species (i.e. the mimic) to an inedible one (i.e. the model). 2. Müllerian mimicry (named after F. Müller (1822–1897), a German-Brazilian naturalist) – an unpalatable or venomous species mimics another such species and both (or more) species have the same predator. Associated with mimicry is flashing colouration in which an animal displays an otherwise obscure or hidden coloured area when disturbed. Such behaviour is seen in certain butterflies, grasshoppers, birds, and ungulates.

(*Continued*)

Table 5.9 (Continued)

Defence mechanism	*Description*
Vocalisation	Some animals vocalise either to warn off predators or as a precursor to fighting. Such calls may be a general, high-pitched noise to which several species respond, or more specific even indicating the type of predator approaching, be it from the air or the ground (e.g. the chicken (*Gallus gallus domesticus*) has a different vocalisation to warm of an aerial predator (e.g. a bird of prey) to that used to warn of a terrestrial predator (e.g. red fox (*Vulpes vulpes*)).
Group living	The number of individuals in a flock may serve to distract a predator from locking on to one individual. Members of a group may come together to mob a predator, thus deterring it from attack.
Chemical defence	An array of chemicals is used defensively by some animals. Pheromones may be released as an alarm and induce fright in members of the same species. Arthropods, amphibians, snakes, and mammals employ secretions, often odorous and produced in copious amounts, to repel predators. Some chemicals are multi-use. For example, venomous snakes use venom both to catch prey and to defend against predators.

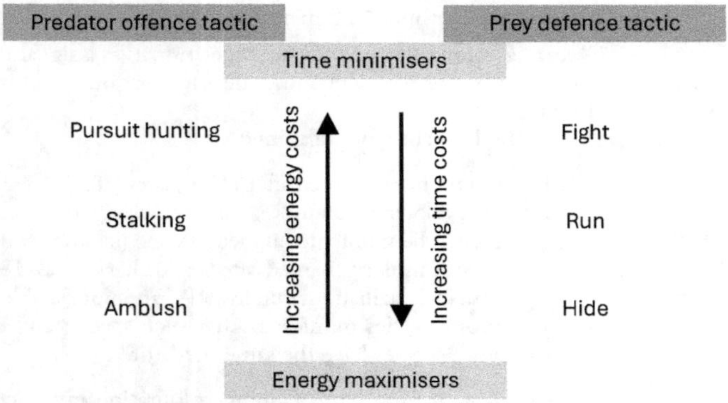

Figure 5.31 Carnivore predator offence tactics and prey defence tactics.

5.7 DO ECOLOGICAL COMMUNITIES EXIST?

A community is a concept that helps ecologists bring order to the natural world in a way that allows the sharing of knowledge and understating about the functions of groups of organisms. However, an often-raised question is 'are communities real?' Several different approaches have emerged to address this question. One approach comes from the argument that species adapt to cope with both the physical/abiotic aspects of their environment and with the presence of other species (see Chapter 3). Species that directly compete adapt to share the resource such that it is partitioned between them. This results in species tolerating each other and, in some cases, becoming dependent on each other. Through these processes of adaptation an assemblage of species can occupy the same habitat in a sustainable way.

For each environmental variable, every species within an assemblage has its own optimal range (see Chapter 3). Plants provide a simple example of this. Plants adapt to factors including soil moisture and acidity, light intensity, and the concentration of individual nutrients. The question is then, do the requirements of groups of species coincide such that they can only occur together in a community, or is the distribution of each species individualistic and independent of each other species and there are no real communities. The first idea, that species requirements coincide to create communities is attributed to Frederic Clements (1874–1945, an American botanist, taxonomist, and ecologist who influenced the early study of plant communities, particularly the process of plant succession) and the second idea, that species act individualistically to Henry Gleason (1882–1975, an American ecologist, botanist, and taxonomist) (Figure 5.32).

Clements viewed a community as an organic entity in which positive interactions and interdependencies lead to distinct associations that are repeated frequently in different locations. This idea was influential in the development of the concept of a community. The question is, however, is this how nature works?

Gleason saw nature as a continuum of environmental variables that constantly change from one location to another. Each species

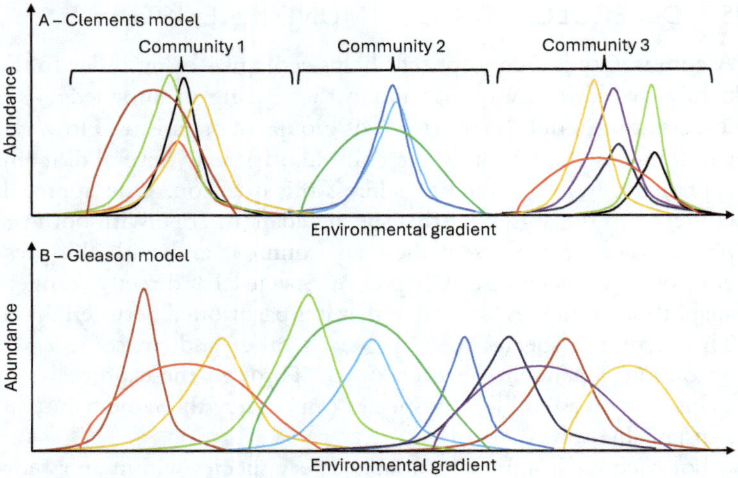

Figure 5.32 Diagrammatic representation of (A) the Clements model of vegetation assemblage in which species requirements coincide leading to distinct communities (three are illustrated) and (B) the Gleason model in which each species is distributed independently and there are no clear communities.

has individualistic requirements. Seen in this way, groups of organisms occur in the same place due to chance and vegetation is a continuum across both space and time. This idea can best be observed in places where there has been no human interference and where environmental variables change gradually. What difference there is has occurred due to disturbances which occur over different scales of time and space (Figure 5.33).

The concept of a community, whether they are the 'superorganism' idea, with predictable successional stages, as put forward by Clements, or the result of individualistic plant adaptations to the environment as proposed by Gleason, is useful in helping make sense of the world around us and in enabling communication between different people.

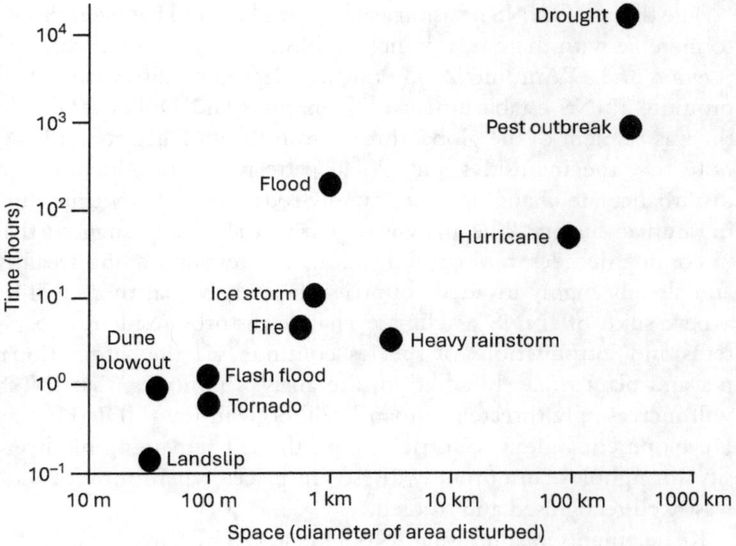

Figure 5.33 Disturbances which lead to unevenness of species.
Source: based on Forman (1995)

5.8 CASE STUDY – INVASIVE NON-NATIVE SPECIES

Species that have spread naturally or have been introduced accidentally or deliberately into a given area since the Neolithic period (which ended around 4,000 years ago) are considered to be native to that area (IUCN 1987). The terms 'non-native species' (NNS), 'alien species' and 'non-indigenous species' are used in various notational and international documents to refer to such species. A subset of NNS is 'invasive non-native species' (INNS): species that have a negative impact on native biodiversity, ecosystem services (see Chapter 6), human well-being, or the economy at various scales. Globally, they are one of the biggest causes of biodiversity loss and species extinctions and are a threat to food security and livelihoods (IUCN 2021).

The threat of INNS invasion at any given location has been shown to increase with the rate at which INNS propagules are introduced (Levine and D'Antonio 2003) and the degree of disturbance that promote INNS establishment (Theoharides and Dukes 2007). In their assessment of the global threats from INNS Early et al. (2016) note that the intensities and global patterns of introduction and disturbance are changing more rapidly today than at any time during human history. The analysis by Early et al. (2016) suggests that in coming decades, biological invasions will remain high in wealthy and already highly invaded countries. It is likely that there will be a new suite of INNS as climate change disturbs resident ecosystems and introductions of species continue via the international pet and plant trade. In addition, the analysis indicates, that INNS will increasingly threaten human livelihoods in low-HDI (Human Development Index) countries and in the last remaining biodiversity strongholds. Importantly, these are places where invasions are least well recognised and studied.

Key elements in addressing INNS identified by Early et al. (2016) include:

- **Prevention**: to prevent entry and reduce establishments of INNS. This requires surveillance at points of entry into a country and more widely across a territory, early detection, and monitoring as well as management to eradicate, control or contain INNS.
- **Evidence**: to ensure that the strategy is based on the best available evidence and identify gaps and priority areas for further research and development.
- **Awareness raising**: to increase awareness of INNS issues and promote appropriate changes in behaviour or attitudes throughout all relevant sectors and among the public.
- **Coordination**: to improve coordination of actions within governments, government associated bodies, and key actors outside government.

There are examples of conservation management strategies and interventions based on ecological knowledge and understanding that may be used to control the spread of INNS (Table 5.10). Despite these varied approaches it is extremely difficult to bring about the total eradication of an INNS with successful eradications limited to small geographic areas, such as the eradication of Norway rats (*Rattus norvegicus*) from the Hebridean island of Canna, UK (Luxmoore et al. 2019), and the eradication of stoat (*Mustela erminea*) across the Orkney archipelago, UK (Bambini et al. 2018).

Strategies, such as those shown in Table 5.10 can be successful. Cullen et al. (2025) report how actions including direct threat

Table 5.10 Examples of strategies and interventions for managing invasive alien species (INNS) that focus on impact-based applications

Problem	*Scale*	*Strategies and interventions*
INNS plants causing reduction in water availability for human use	Country level	INNS control
Offsetting of INNS impacts	Landscape and patch level (islands and sanctuaries)	Create insurance populations by translocation of the species or asset affected to an INNS-free place.
Impacts of plant INNS	Landscape and local levels	Active restoration or promotion of natural regeneration of native plant communities by restoring diverse native plant communities on disturbed or degraded sites.
INNS agricultural pests.	Local level (properties and plantations)	Creation and plantation of crop varieties that are resistant to agricultural pests.
Effects of mammalian INNS on livestock, poultry, nesting grounds, and insurance populations	Local level (properties, paddocks, and patches)	Install and maintain fencing to exclude mammalian INNS.
Protection from mammal INNS predators.	Local level (properties and paddocks)	Use dogs to protect poultry, livestock, and biodiversity from mammalian INNS predators.
INNS agricultural bird and insect pests eating and damaging of food production crops	Very local level (e.g. orchards within properties or individual trees)	Use nets and fences to cover and exclude INNS insects and birds from crops and fruit production trees.

(*Continued*)

Table 5.10 (Continued)

Problem	*Scale*	*Strategies and interventions*
Predation of individual nests of birds and turtles.	Very local level (prey population and individual nests)	Use individual nest cages excluding predators (including INNS).
Impacts of an INNS prey on a native predator.	Local level (predator populations)	Use of toxic sausages to train quolls (*Dasyurus hallucatus*) to avoid eating the poisonous, toxic invasive cane toads (*Rhinella marina*)
INNS predation pressure on vulnerable native bird nests	Local and very local levels (prey populations and individual nest levels)	Use of compounds to conceal or deter INNS predators (e.g. rat (*Rattus* spp.) and hedgehog (*Erinaceus europaeus*)) from bird nests.

Source: based on García-Díaz et al. (2021)

abatement; invasive predator control, and construction of safe havens (feral predator and herbivore-free fenced conservation reserves), translocation of species, habitat restoration, conditioning predator recognition/aversion, and captive breeding programmes are being used in Australia to promote the recovery of native mammal species populations. The mammal species most at risk are within a critical weight range (CWR) between 35 g and 5.5 kg. These are species that are predated by introduced cats (*Felis catus*) and red fox (*Vulpes vulpes*). Such predation is the main cause of their death and the decline in the populations of the prey species is compounded by competition from and habitat degradation by rabbits (*Oryctolagus cuniculus*). The rabbit was introduced from Europe to south-eastern Australia in the 1860s and is a prey species for cats and red fox. The introduction of biocontrol measures (rabbit haemorrhagic disease virus (RHDV)) to reduce the rabbit population led to a decrease in the populations of cats and red fox and an increase in native rodent and invertebrates. The latter are prey items of predatory species in the CWR such as the crest-tailed mulgara or ampurta (*Dasycercus hillieri*). With the changes in the community resulting from the control of rabbits (which reduced predator pressure and resulted in there being more prey species), the range of the ampurta was reported to have increased by 70-fold between 1995 and 2015 (Pedler et al. 2016).

5.9 SELF-STUDY SUGGESTIONS

5.9.1 SELF-STUDY QUESTIONS

1. What is a community?
2. What is the vertical structure of aquatic and terrestrial communities?
3. What is the biological structure of a community?
4. Distinguish between species dominance, species richness, and species diversity. How are these concepts interrelated?
5. What are the typical layers found within the vertical structure of a forest?
6. What are the different elements typically found in the horizontal structure of a stream or river?
7. What are the differences and similarities between the Grinnellian, Eltonian, Hutchinsonian, and contemporary concepts of a niche?
8. How does the concept of niche separation help to explain why many species can live in the same area at the same time?
9. What are the differences between a fundamental and realised niche?
10. What are the main trophic levels in ecosystems and how do these interact?
11. What are the main differences between a grazing and detrital food chain?
12. How do grazing and detrital food chains interact?
13. What are the differences between a food chain and a food web and how are they interrelated?
14. What are the differences between primary and secondary succession?
15. What are the distinguishing features of early successional plants?
16. What abiotic and biotic factors might influence the rate of habitat recovery during secondary succession?
17. Explain the concept of a sere and give an example of a hydrosere and a halosere.
18. What are pyramids of numbers and pyramids of biomass?
19. Why can pyramids of numbers and pyramids of biomass sometimes be inverted?
20. What do these words mean in an ecological sense: mutualism, commensalism, amensalism, parasitism, and predation?
21. What defence mechanisms are used by plants against herbivores?

22. What defence mechanisms are used by animals against carnivores?
23. What are the three main tactics used by predators when hunting prey and what are the advantages and disadvantages of each tactic?
24. Compare Clements's and Gleason's ideas about communities?
25. What are invasive species, and what effects do they have on a community?

5.9.2 A DEEPER DIVE

A. There are various databases of home range available (e.g. HomeRange, https://shoeks.github.io/HomeRange). Explore such a database and look up the home ranges of animals that you are familiar with.
B. Explore the Global Invasive species database (GISD) www.iucngisd.org/gisd/ and find invasive species that are relevant to where you live and/or study.
C. What is the 'arms race' between predators and prey?

5.10 REFERENCES

Angrish, R. (2022). 'The wood wide web'. *Resonance* 27: 1429–1441. https://doi.org/10.1007/s12045-022-1435-x.

Bambini, L., Sankey, S. and Neville, G. (2018). '*Orkney Native Wildlife Project: Stoat Eradication Operational Plan*'. RSPB Scotland unpublished report. Retrieved from www.nature.scot/sites/default/files/2020-05/Orkney%20Native%20Wildlife%20Project%20-%20SEA%20consultation%20-%20Operational%20plan.pdf (accessed 6 August 2025).

Bazzaz, F.A. (1979). 'The physiological ecology of plant succession'. *Annual Review of Ecology, Evolution, and Systematic* 10: 351–371. www.jstor.org/stable/2096796.

Brattsten, L.B., Wilkinson, C. F., and Eisner, T. (1977). 'Herbivore-plant interactions: mixed-function oxidases and secondary plant substances'. *Science.* 196: 1349–1352. https://doi.org/10.1126/science.196.4296.1349.

Buzhdygan, O.Y., et al. (2020). 'Biodiversity increases multitrophic energy use efficiency, flow and storage in grasslands'. *Nature Ecology and Evolution* 4: 393–405. https://doi.org/10.1038/s41559-020-1123-8.

Castro-Delgado, A.L., Elizondo-Mesén, S., Valladares-Cruz, Y. and Rivera-Méndez, W. (2020). 'Wood Wide Web: communication through the mycorrhizal network', 'Internet de las plantas: comunicación a través de la red micorrízica'. *Tecnología en Marcha* 33: 4.

Chase, J.M. and Leibold, M.A. (2003). *Ecological Niches: Linking Classical and Contemporary Approaches*. Chicago, IL: University of Chicago Press.

Cullen, D., Kingsford, R.T., Bino, G., West, R., Letnic, M. and Pedler, R. (2025). 'Bucking the trend – recovery from near continent-wide extinction by a marsupial micro-predator during drought'. *Biological Conservation* 311: 111411. https://doi.org/10.1016/j.biocon.2025.111411.

Dayton, P.K. (1972). 'Toward an understanding of community resilience and the potential effects of enrichments to the benthos at McMurdo Sound, Antarctica'. In B.C. Parker (ed.), *Proceedings of the Colloquium on Conservation Problems in Antarctica,* 81–96. Allen Press.

Dussourd, D.E. and Denno, R.F. (1991). 'Deactivation of plant defense: correspondence between insect behaviour and secretory canal architecture'. *Ecology* 72: 1383–1396. https://doi.org/10.2307/1941110.

Dussourd, D.E. and Eisner, T. (1987). 'Vein-cutting behaviour: insect counterploy to the latex defense of plants'. *Science* 237: 898–901. https://doi.org/10.1126/science.3616620.

Early, R., et al. (2016). 'Global threats from invasive alien species in the twenty-first century and national response capacities'. *Nature Communications* 7: 12485. https://doi.org/10.1038/ncomms12485.

Elhakeem, A., Markovic, D., Broberg, A., Anten, N.P.R. and Ninkovic, V. (2018). 'Aboveground mechanical stimuli affect belowground plant-plant communication'. *PLOS ONE* 13(5): e0195646. https://doi.org/10.1371/journal.pone.0195646.

Ellison, A.M., et al. (2005). 'Loss of foundation species: consequences for the structure and dynamics of forested ecosystems'. *Frontiers in Ecology and the Environment* 3(9): 479–486. https://doi.org/10.1890/1540-9295(2005)003[0479:LOFSCF]2.0.CO;2.

Elton, C. (1927). *Animal Ecology*. London: Sidgwick and Jackson.

Emery, S.M. (2010). 'Succession: a closer look'. *Nature Education Knowledge* 3(10): 45.

Forman, R.T.T. (1995). *Land Mosaics: The Ecology of Landscapes and Regions*. Cambridge: Cambridge University Press.

García-Díaz, P., et al. (2021). 'Management policies for invasive alien species: addressing the impacts rather than the species'. *BioScience* 71(2): 174–185. https://doi.org/10.1093/biosci/biaa139.

Gilbert, L.E. (1975). 'Ecological consequences of a coevolved mutualism between butterflies and plants'. In: L.E. Gilbert and P.H. Raven (eds), *Coevolution of Animals and Plants*, 210–240. Austin, TX: University of Texas Press.

Gorzelak, M.A., Asay, A.K., Pickles, B.J. and Simard, S.W. (2015). 'Inter-plant communication through mycorrhizal networks mediates complex adaptive behaviour in plant communities'. *AoB PLANTS* 7: plv050. https://doi.org/10.1093/aobpla/plv050.

Grinnell, J. (1917). 'The niche-relationships of the California Thrasher'. *The Auk* 34(4): 427–433. https://doi.org/10.2307/4072271.

Hill, J.E., DeVault, T.L. and Belant, J.L. (2019). 'Cause-specific mortality of the world's terrestrial vertebrates'. *Global Ecology and Biogeography* 28(5): 680–689. https://doi.org/10.1111/geb.12881.

Hutchinson, G. E. (1957). 'Concluding remarks'. *Cold Spring Harbor Symposium on Quantitative Biology* 22: 415–427. https://doi.org/10.1101/SQB.1957.022.01.039.

IUCN. (1987). *The IUCN Position Statement on Translocation of Living Organisms.* Gland, Switzerland: IUCN.

IUCN. (2021) *Annual Report*. Gland, Switzerland: IUCN.

Janzen, D.H. (1971). 'Seed predation by animals'. *Annual Review of Ecology, Evolution, and Systematics* 2: 465–492. https://doi.org/10.1146/annurev.es.02.110171.002341.

Letten, A.D., Ke, P-J and Fukami, T. (2017). 'Linking modern coexistence theory and contemporary niche theory'. *Ecological Monographs* 87(2): 161–177. https://doi.org/10.1002/ecm.1242.

Levine, J.M. and D'Antonio, C.M. (2003). 'Forecasting biological invasions with increasing international trade'. *Conservation Biology* 17: 322–326. https://doi.org/10.1046/j.1523-1739.2003.02038.x.

Luxmoore, R., Swann, R. and Bell, E. (2019). 'Canna seabird recovery project: 10 years on'. In C.R. Veitch, M.N. Clout, A.R. Martin, J.C. Russell and C.J. West (eds), *Island Invasives: Scaling Up to Meet the Challenge*, 576–579. Occasional Paper SSC no. 62. Gland, Switzerland: IUCN.

Moran, N.A. (2006). 'Symbiosis'. *Current Biology* 16(20): 866–871. https://doi.org/10.1016/j.cub.2006.09.019.

Paine, R.T. (1966). 'Food web complexity and species diversity'. *The American Naturalist* 100: 910 https://doi.org/10.1086/282400.

Paine, R.T. (1969). 'The Piaster-Tegula interaction: prey patches, predator food preference, and intertidal community structure'. *Ecology* 50(6): 950–961. https://doi.org/10.2307/1936888.

Paracer, S. and Ahmadjian, V. (2000). *Symbiosis: An Introduction to Biological Associations*. Oxford: Oxford University Press.

Pedler, R.D., Brandle, R., Read, J.L., Southgate, R., Bird, P. and Moseby, K.E. (2016). 'Rabbit biocontrol and landscape-scale recovery of threatened desert mammals'. *Conservation Biology* 30(4): 774–782. https://doi.org/10.1111/cobi.12684.

Pielou, E.C. (1966). 'The measurement of diversity in different types of biological collections'. *Journal of Theoretical Biology* 13: 131–144. https://doi.org/10.1016/0022-5193(66)90013-0.

Pocheville, A. (2015). 'The ecological niche: history and recent controversies'. In T. Heams, P. Huneman, G. Lecointre, and M. Silberstein (eds), *Handbook of*

Evolutionary Thinking in the Sciences. Dordrecht: Springer. https://doi.org/10.1007/978-94-017-9014-7_26.

Saeedghalati, M. (2017). 'Quantitative comparison of abundance structures of generalized communities: from B-cell receptor repertoires to microbiomes'. *PLOS Computational Biology* 13(1): e1005362. https://doi.org/10.1371/journal.pcbi.1005362.

Shannon, C.E. (1948). A mathematical theory of communication'. *Bell System Technical Journal* 27: 379–423. https://doi.org/10.1002/j.1538-7305.1948.tb01338.x.

Simard, S.W., Perry, D.A., Jones, M.D., Myrold, D.D., Durall, D.M. and Molina, R. (1997). 'Net transfer of carbon between tree species with shared ectomycorrhizal fungi'. *Nature* 388: 579–582. https://doi.org/10.1038/41557.

Simpson, E.H. (1949) 'Measurement of diversity'. *Nature* 16: 688.

Takola, E. and Schielzeth, H. (2022). 'Hutchinson's ecological niche for individuals'. *Biology & Philosophy* 37: 25. https://doi.org/10.1007/s10539-022-09849-y.

Theoharides, K.A. and Dukes, J.S. (2007). 'Plant invasion across space and time: factors affecting nonindigenous species success during four stages of invasion'. *New Phytologist* 176: 256–273. https://doi.org/10.1111/j.1469-8137.2007.02207.x.

Whittaker, R.H. (1965). 'Dominance and diversity in land plant communities: numerical relations of species express the importance of competition in community function and evolution'. *Science* 147(3655): 250–260. https://doi.org/10.1126/science.147.3655.250.

Whittaker, R.H. (1972). 'Evolution and the measurement of species diversity'. *Taxon* 21: 213–251.

Wilcox, B.A. (1984). 'In situ conservation of genetic resources: determinants of minimum area requirements'. In J.A. McNeely and K.R. Miller (eds), *National Parks, Conservation and Development, Proceedings of the World Congress on National Parks*, 18–30. Washington, DC: Smithsonian Institution Press. https://doi.org/10.13140/2.1.4879.2322.

Williams, K.S. and Gilbert, L.E. (1981). 'Insects as selective agents on plant vegetative morphology: Egg mimicry reduces egg laying by butterflies'. *Science* 212: 467–469. https://doi.org/10.1126/science.212.4493.467.

Zalucki, M.P. and Malcolm, S.B. (1999). 'Plant latex and first-instar monarch larval growth and survival on three North American milkweed species'. *Journal of Chemical Ecology* 25: 1827–1842. https://doi.org/10.1023/A:1020929732223.

6

ECOSYSTEM ECOLOGY

6.1 INTRODUCTION

The concept of an ecosystem combines the abiotic and biotic components of the environment into a unified, functional unit. Ecosystem ecologists examine how ecosystems function (i.e. how energy flows through and material cycle around an ecosystem). They also study the movement of species into and out of an ecosystem. Ecosystem ecologists seek to understand the roles different species play within these systems and how ecosystems respond to natural and human caused changes. To do so, they frequently draw together an understanding of the abiotic components of natural systems (Chapter 2) with that of various levels of biological organisation including individual organisms (Chapter 3), populations (Chapter 4), communities (Chapter 5), ecosystems (this chapter) and landscapes (Chapter 7).

The first section of this chapter contains an overview of the concept of an ecosystem, of ecosystem processes, and of associated emergent ecosystem properties. The second section contains an examination of the flow of energy through an ecosystem (an ecosystem process). The processes of photosynthesis and chemosynthesis are described along with productivity and energy transfer between trophic levels.

The focus in the third section is on another ecosystem process – the cycling of materials within an ecosystem (aka biogeochemical cycles) including that of novel compounds such as plastics. The processes of biomagnification, bioaccumulation, bioconcentration, and biodilution are discussed in this section.

In the fourth section the theme is non-equilibrium systems. Their importance is explained and their implications are examined

DOI: 10.4324/9781003604464-6

with a focus on the stability and resilience (both emergent properties of ecosystems). In the next section, the concept of an ecosystem is broadened to that of a social-ecological system. This emphasises the importance of humans as part of, rather than being distinct from the natural processes occurring on earth.

Ecosystems provide essential services that support human life and well-being, known as ecosystem services, including the provision of food, water, oxygen, timber, and medical resources and the regulation of climate, water purification, and pollination of crops. They also offer opportunities for recreation, aesthetic enjoyment, and spiritual benefits from contact with and an appreciation of nature. Ecosystems that are composed of species rich communities and diverse abiotic factors can withstand environmental changes and disturbances, such as climate change or habitat destruction better than those ecosystems which have been simplified by natural processes or human activity. This highlights the contribution of biodiversity to the resilience and stability of ecosystems. Ecosystem ecologists provide insights into the impacts of human activities on natural systems, help develop strategies for conservation and restoration, and contribute to addressing global challenges such as biodiversity loss, climate change, and pollution.

The focus of the case study in this chapter is nature-based solutions (NbS). Nature-based solutions are based on ecological understanding of ecosystems and aim to restabilise ecosystems so that they are functional and can deliver benefits to people.

6.2 ECOSYSTEMS

A.G. Tansley (1871–1955, a British botanist and pioneer of the science of ecology) believed that the concept of a community (see Chapter 5) was incomplete as it only included the biotic components of the environment. His premise was that biological and abiotic components (see Chapter 2) interact and should be considered together when studying the natural environment. In 1935, he brought together abiotic and biotic components into a functioning unit which he called an ecosystem (Tansley 1935) (Figure 6.1).

An ecosystem is an open system (rather than a closed system with hard boundaries that nothing crosses) with energy, materials, and organisms flowing in and out through porous boundaries. Energy

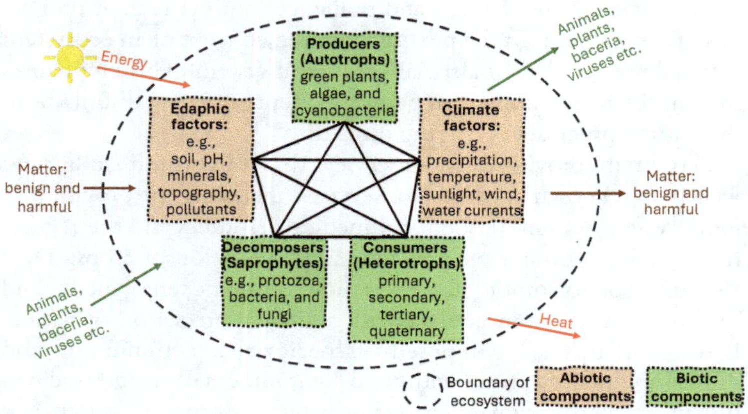

Figure 6.1 Components of an ecosystem.

enters most ecosystems as sunlight and is lost as heat. At a global scale, the Earth receives some matter in the form of meteorites while molecules of gas are lost from the atmosphere to space. These loses and gains are small in comparison to the entire Earth. On a smaller geographical scale, material is eroded by wind and water and transported from one location, and from one ecosystem to another. The third important aspect of this porosity is that microorganisms, plants, and animals move into and out of an ecosystem (as adults, juveniles, and seeds and spores according to the species) either because they are mobile or as a result of being transported by air, water, or animals (including people) (Figure 6.2).

As with the concept of a community, the boundaries of an ecosystem are only fixed by those studying it. It is, therefore, possible to treat the entire Earth as an ecosystem. Equally, a lake, a coral reef, a hydrothermal vent, a forest, an area of grassland, a suburban garden, a single tree, or a clump of grass can all be viewed as ecosystems (Figure 6.3).

A modern definition of an ecosystem is provided by the UN's Convention on Biological Diversity – 'a dynamic complex of plant, animal, and micro-organism communities and their non-living environment interacting as a functional unit' (United Nations 2011, p. 4). This definition explicitly mentions function. Functions are the underpinning processes of an ecosystem and are frequently termed ecosystem processes. These processes include energy flow,

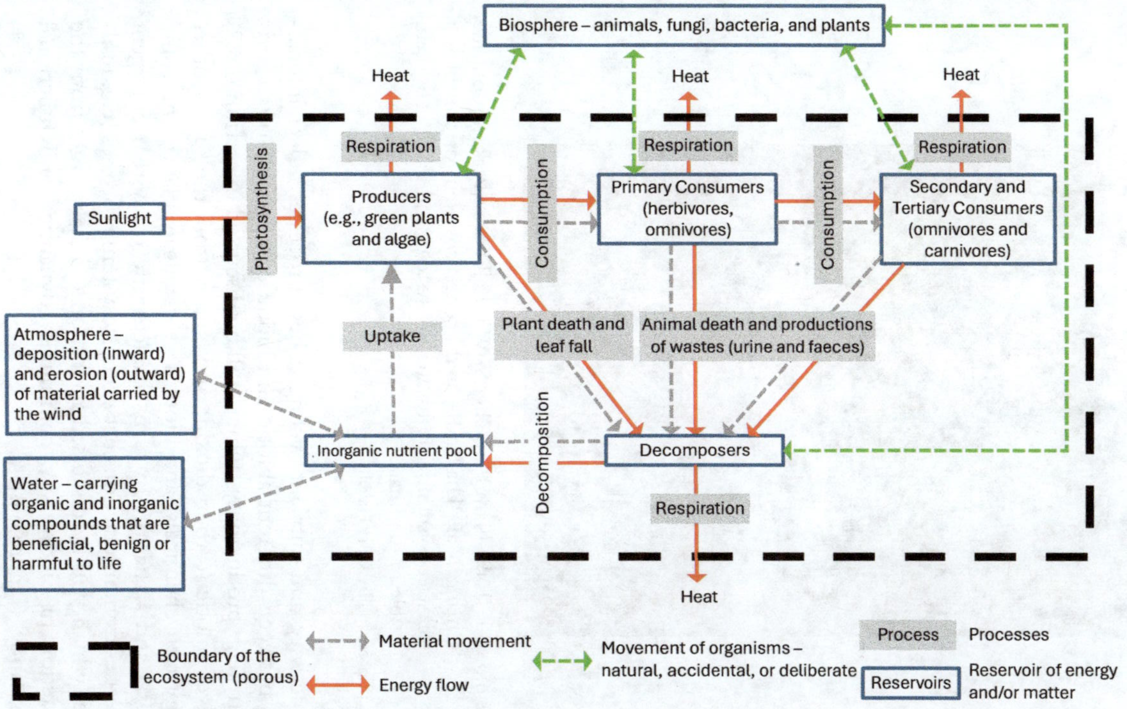

Figure 6.2 Schematic model of an ecosystem illustrating the flow of energy through and cycling of matter into around an ecosystem.

Figure 6.3 Bua Tong-Nam Phu Chet Si Waterfall Forest Park, just outside Chiang Mai, Thailand – is this one ecosystem or are there a forest ecosystem, a freshwater ecosystem, and a soil ecosystem visible in the image?

Source: photograph by Gina Hazell

productivity, and nutrient cycling. These processes reflect interactions between different abiotic and biotic components of an ecosystem (e.g. productivity (which will be discussed in section 6.3.2) involves abiotic (carbon dioxide and energy from the sun or other energy rich chemicals) and biotic factors (algae, plants, and some bacteria and archaea)). In turn characteristic ecosystem emergent properties from interactions between ecosystem processes (e.g. resilience, which will be discussed in section 6.5.2, emerges from the interaction of biodiversity, community dynamics, and functional redundancy) (Figure 6.4).

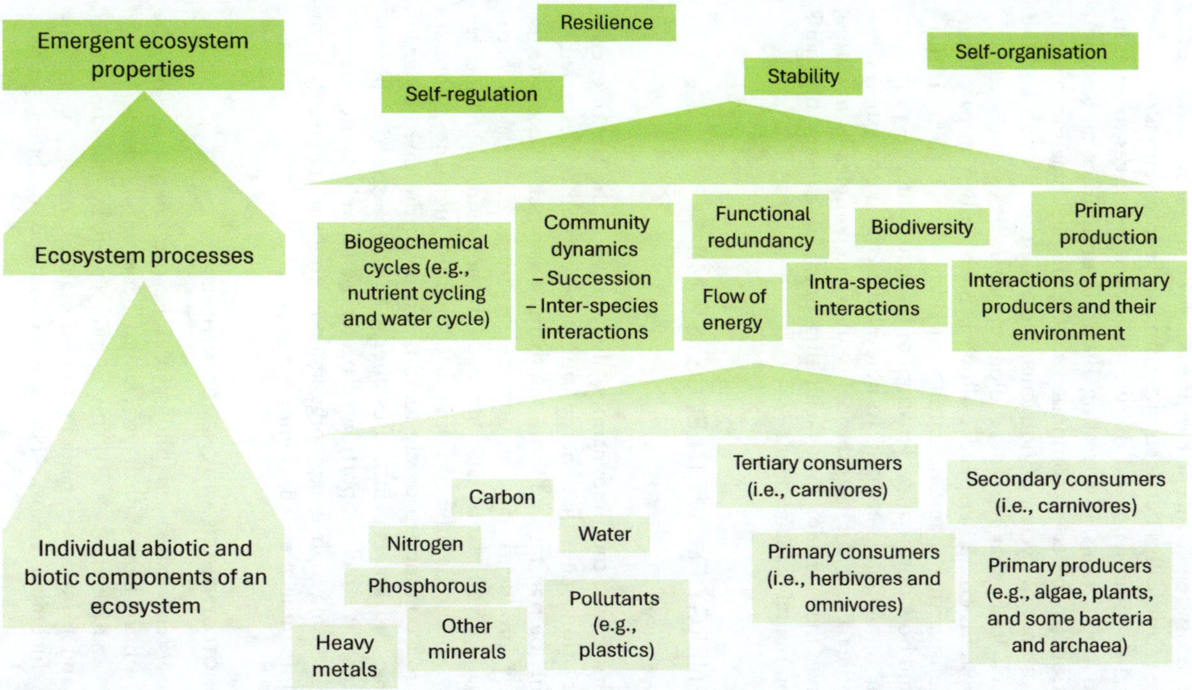

Figure 6.4 Examples of ecosystem components, processes, and emergent properties (in some discussions some ecosystem processes (e.g. energy flow and biogeochemical cycles) are described as emergent properties).

The ecosystem concept has proved to be useful in understanding the relationships between organisms, their interactions with the physical environment, and the use of natural resources to support human populations. By treating biotic communities and abiotic conditions together as components of a single, dynamic system, flows of energy, cycles, of nutrients, movement of species, and the feedback loops that regulate ecological processes can be analysed. This systems-based perspective allows habitats to be compared, the mechanisms that maintain stability or drive change to be identified, and how disturbances – natural or human-induced – propagate through ecological networks to be evaluated. The ecosystem concept offers a unifying lens through which the complexity of nature becomes manageable, thus facilitating practical conservation and environmental management.

6.3 ENERGY FLOWS IN ECOSYSTEMS

6.3.1 OVERVIEW

The first law of thermodynamics, also known as the law of energy conservation, states that energy cannot be created or destroyed, it can only be transformed from one form to another. In ecosystems, primary producers (aka autotrophs i.e. some bacteria, archaea, algae, and green plants) take energy from the environment and use it to synthesise organic compounds containing high-energy bonds. The most common pathway for this synthesis, the one employed by algae and green plants, uses energy from the sun (i.e. photosynthesis). A small proportion of bacteria and archaea are capable of synthesising high-energy organic compounds using energy from inorganic molecules (i.e. chemosynthesis).

6.3.2 PHOTOSYNTHESIS AND CHEMOSYNTHESIS

In ecosystems based on photosynthesis, the energy in sunlight (usually from light with a wavelength between 400 and 700 nm (nm = nanometre i.e. 10^{-9} m) is intercepted by chloroplasts in algae and plant leaves (and in some instances, other green plant tissues including stems). This energy is used to build glucose (a carbohydrate) which has energy rich bonds from carbon dioxide and water. The process of photosynthesis can be summarised as:

$$\text{carbon dioxide} + \text{water} \rightarrow \text{glucose} + \text{oxygen}$$
$$6CO_2 + 6H_2O \rightarrow C_6H_2O_6 + 6O_2$$

The glucose produced is further processed by plants to produce cellulose, starch, nucleic acids, proteins, lipids, and various other organic compounds, which all contain high amounts of chemical energy.

In 1977, the discovery of thriving ecosystems at deep-sea hydrothermal vents where there is no light (i.e. photosynthesis is not possible) changed views of biology (Dick 2019). The water issuing from the vents is mineral rich and contains sulphur compounds, particularly hydrogen sulphide (H_2S). When hydrogen sulphide is oxidised, energy is released. Some of this energy is stored in glucose by chemosynthetic bacteria (Dick 2019; Ramirez-Llodra and Keith 2020). The process of chemosynthesis can be summarised as:

$$\text{carbon dioxide} + \text{hydrogen sulphide} + \text{oxygen}$$
$$\rightarrow \text{glucose} + \text{sulphur} + \text{water}$$
$$CO_2 + 4H_2S + O_2 \rightarrow CH_2O + 4S + 3H_2O \quad \text{(simplified equation)}$$

Hydrothermal vent ecosystems contain animals which obtain the energy they require by feeding on the chemosynthetic bacteria and archaea. The vents are also a significant source of iron for phytoplankton in the oceans and, hence, their impact is far wider than their immediate vicinity (Resing et al. 2015).

6.3.3 PRODUCTIVITY

The energy accumulated by autotrophs is called productivity, or more specifically, primary productivity (primary being a reference to it being the first and most basic form of energy storage in an ecosystem). All the energy assimilated represents the gross primary productivity (GPP). GPP can be affected by environmental factors. Lomax et al. (2024) analysed remote sensing-derived datasets from 2000 to 2019 and quantified the effect of precipitation amount, distribution, and intensity on the GPP of dry rangelands across sub-Saharan Africa. Total precipitation was found to be the primary driver of productivity, but more variable rainfall had a small negative effect across vegetation types and rainfall regimes. They also found that temperature and soil nitrogen have strong effects, especially in drier areas.

All organisms require energy to maintain basic metabolic processes. This energy is released from the energy-rich chemical bonds in glucose during the process of respiration. The process of respiration can be summarised as:

$$\text{glucose} + \text{oxygen} \rightarrow \text{carbon dioxide} + \text{water}$$

$$C_6H_{12}O_6 + 6O_2 \rightarrow 6CO_2 + 6H_2O$$

Most of the released energy is stored in ATP (adenosine triphosphate), which is then used in other metabolic reactions, but some is lost as heat (Figures 6.2 and 6.3). This means that not all the GPP is converted into biomass. The energy that does accumulate is termed net primary productivity. That is to say:

$$\text{net primary productivity}\left(\text{NPP}\right)$$
$$= \text{gross primary productivity}\left(\text{GPP}\right) - \text{respiration}\left(\text{R}\right)$$

The net primary productivity of different ecosystems varies and due to differences in the area of each ecosystem, there are also big differences in global net primary productivity (Table 6.1).

Net primary productivity in an ecosystem varies between years and also at different stages of succession (see Chapter 5) as the ecosystem ages. Year-to-year variations in the physical environment that affect NPP include temperature, rainfall, nutrient availability (particularly of nitrogen and phosphorous), hours of sunshine, herbivory, and disturbances such as fire, flood, and landslips. The effect of the age of an ecosystem on NPP can be seen in forests. Typically, the net primary productivity of a forest increases rapidly during the early stage of succession, reaches a maximum in middle age, and gradually declines in later ages (He et al. 2012). In general, absolute and relative growth rates (i.e. productivity) tend to decrease with the increasing age and height of a tree. Reduced photosynthesis

Table 6.1 Net primary productivity per unit area and global net primary productivity for selected world ecosystems. gC = grams of carbon. Global net primary productivity = mean net primary productivity per unit area × global area of the relevant ecosystem

Ecosystems (in order of mean net primary productivity)	*Mean net primary productivity per unit area ($gC\ m^{-2}\ yr^{-1}$)*	*Global net primary productivity ($10^9\ gC\ yr^{-1}$)*
Aquatic		
Algal beds and reefs	2000.0	1.10
Estuaries	1800.0	2.40
Lake and stream	500.0	1.30
Upwelling zones	500.0	0.22
Continental shelf	360.0	9.60
Open ocean	127.0	42.00
Terrestrial		
Swamp and marsh	2500.0	4.90
Tropical rain forest	2000.0	34.00
Tropical seasonal forest	1500.0	11.30
Temperate evergreen forest	1300.0	6.40
Temperate deciduous forest	1200.0	8.40
Boreal forest	800.0	9.50
Savanna	700.0	10.40
Cultivated land	644.0	9.10
Woodland and shrubland	600.0	4.90
Temperate grassland	500.0	4.40
Tundra and alpine meadow	144.0	1.10
Desert shrub	71.0	1.30
Rock, ice, sand	3.3	0.09

Source: adapted from Whittaker and Likens (1973)

with plant age, decreased nutrient supply with stand age (i.e. the time that trees have been growing in that place), and increased reproductive effort and increased mortality of older individuals all reduce net primary productivity as the relative balance between photosynthesis and respiration change over time (Ryan et al. 1997; Kutsch et al. 2009).

6.3.4 ENERGY TRANSFER BETWEEN TROPHIC LEVELS

The net energy gained by a herbivore eating a plant or a carnivore eating an animal is termed secondary productivity. It is the energy consumed minus that which is passed out of the body as waste products (faeces, urine, and fermentation gases (e.g. methane (CH_4))) and that which is lost as heat used to maintain resting body temperature in homeothermic (aka homothermic) animals. This net energy can be used for maintenance, growth, and reproduction. This includes the energy used during the animal's daily routine, in harvesting or capturing food, and that needed to repair the wear and tear on the animal's body (Figure 6.5).

Secondary production depends on the availability quantity and quality (including nutritional status and digestibility) of the food source. Hence, any constraints on the food source affects secondary production. For example, Brylinsky and Mann (1973) demonstrated that, within a growing season, zooplankton production was directly related to phytoplankton production (i.e. more food (phytoplankton) results in more zooplankton) and Coe et al. (1976) showed that large herbivore productivity in Africa was directly related to

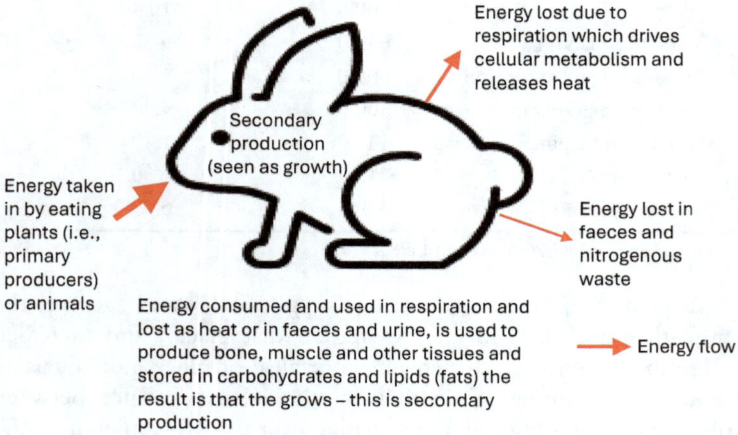

Figure 6.5 Secondary productivity.

rainfall (i.e. more rainfall, more plant material, and, therefore, more secondary productivity in the large herbivores).

The ratio of assimilation to consumption or ingestion (A/I, where A = energy assimilated and I = energy ingested) is a measure of the efficiency of the consumer at extracting energy from the food it consumes. The ratio of productivity to assimilation (P/A, where P = secondary productivity and A = energy assimilated) is called productivity efficiency and is a measure of the efficiency of a consumer in incorporating assimilated energy into new tissue (i.e. growth), or secondary productivity. A wide range of conversion efficiencies exist among different feeding groups. In plants the ratio of net productivity to light absorbed is low, ranging between 0.34 per cent in some phytoplankton to 0.8 to 0.9 per cent in grassland vegetation. Herbivore efficiency ranges from 5–16 per cent and the range for carnivores is around 13–24 per cent (Andrzejewska and Gyllenberg 1980). These figures indicate a relative low efficiency in converting energy obtained from one trophic level into productivity in the next tropic level. Due to this relative inefficiency, most food chains are no longer than three or four and very rarely five levels (Pimm 1982). This explains why large carnivores are relatively rare (see also Colinvaux 1978).

6.4 BIOGEOCHEMICAL CYCLES

6.4.1 MATTER

As stars of a certain size die, they collapse and become extremely hot and dense and explode. These conditions are sufficient for nuclear fusion to create all the naturally occurring chemical elements that we know. When a star explodes, the chemical elements it contained are scattered into a cloud of star dust. Around 4.6 billion years ago, within such a cloud of star dust, the Earth began to form as dust particles collided and stuck together in ever larger aggregates (Sutter 2023). All the matter currently found on earth came from that dust cloud, apart from some relatively small additions from meteorites that fall to earth.

All chemical elements on earth cycle through various reservoirs, sometimes as the element itself and at other times as part of a molecule. The main reservoirs are the atmosphere (the blanket of gases that surrounds the Earth), the lithosphere (the rigid outer part of the earth, consisting of the crust and upper mantle), the hydrosphere (the layer of water at or near Earth's surface which includes all liquid and frozen surface waters, groundwater held in soil and rock, and atmospheric water vapour), and the biosphere (the parts of the earth where life exists). There are numerous mechanisms which cause movement of materials between these reservoirs including oceanic gas exchange, photosynthesis, respiration, decomposition, combustion, rock weathering, volcanism, and industrial processes. Biogeochemical cycles are the pathways by which atoms and molecules flow between the reservoirs (Figure 6.6).

The movement of elements through their cycles can sometimes be fast and, at other times, extremely slow. Fast cycles are complete over years or decades. Slow cycles can take millions of years. An example of a fast cycle is the absorption of carbon dioxide into plants and algae during photosynthesis and the subsequent release of some of that carbon dioxide back into the atmosphere from respiration. An example of a slow cycle is the movement of carbon between the atmosphere, lithosphere, and oceans which can take many tens of thousands of years. The fast and slow cycles can interact with each other in feedback loops.

In addition to the cycling of naturally occurring elements and molecules, chemicals produced synthetically which enter ecosystems also have fast or slow cycles depending on the chemical. Examples of such chemicals include heavy metals, plastics, pesticides, radioactive isotopes, artificial nutrients (e.g. plant fertilisers), and several toxic substances. While some chemicals are benign, others disrupt the functioning of ecosystems and, in some cases due to their toxicity to specific organisms produce sub-lethal or lethal effects. Some chemicals break down in air, water, soil, or are metabolised to less harmful compounds by organisms, while others build up in reservoirs.

Plastics are used in almost every aspect of daily life due to their lightness, high durability, ductility, strength, and cost-effectiveness (Tran et al. 2022). All plastics are manufactured from hydrocarbons and together the various types of plastic produced contain more than 10,000 chemicals, including carcinogens and endocrine disruptors. Most forms of plastic have a long half-life meaning that once they enter ecosystems, they persist for a considerable time (Chamas et al. 2020) (Figure 6.7).

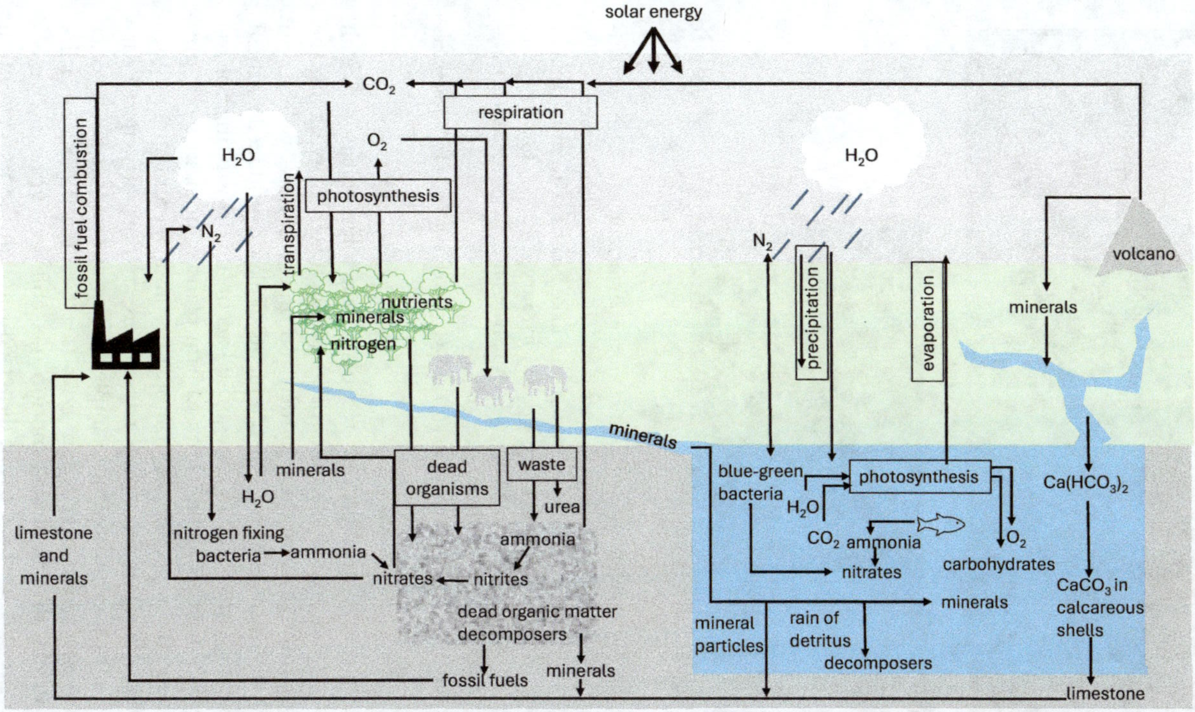

Figure 6.6 Flows of the major biological elements (carbon and nitrogen), minerals, and water around an ecosystem.

Source: based on Hutchinson (1970)

Figure 6.7 The persistence of plastics in the environment. (A) Nest constructed by village weaver birds (*Ploceus cucullatus*) using plastic cords and natural materials near Lake Naivasha, Kenya. (B) Nest of a mute swam (*Cygnus olor*) in Bury, UK. (C) On the water surface with feeding mallard female and ducklings (*Anas platyrhynchos*) in Manchester, UK.

Sources: (A) photograph by Rob Young; (B) and (C) photographs by Luke Blazejewski

Plastics may degrade due to solar ultraviolet radiation which induces photooxidation, microbial activity which weakens the structural integrity of the polymer, and, in the oceans, wave action which exerts shear stress. The fragments of plastic produced by these degradation processes are categorised according to size: macroplastic > 5 mm, microplastics 1 μm to 5 mm, and nanoplastics < 1 μm. In water the dispersion of macroplastics and microplastics is governed by their buoyancy properties. There are hotspots where floating plastics accumulate, for example, in bays and convergence zones such as the subtropical ocean gyres (see Chapter 2). Nanoplastics are so small that Brownian motion (i.e. collision with water molecules) results in them being dispersed through the water column (ten Hietbrink et al. 2025). Plastics enter water courses by various routes including runoff from roads, sewage, and grey water from washing machines. In terrestrial environments particles of all sizes are transported by winds from one location to another.

In addition to their physical presence, micro- and nanoplastics are vectors for many pollutants such as heavy metals, polyaromatic hydrocarbons (PAHs), polychlorinated biphenyls (PCBs), and other toxic pollutants (Guo and Wang 2019; Gaur et al. 2021a, 2021b; Varjani et al. 2020). These chemicals can be released throughout the entire life cycle of the plastic – including feedstock extraction (i.e. crude oil extraction from below ground), production, use, and waste disposal. Certain waste disposal methods, such as uncontrolled landfilling or incineration, can further exacerbate chemical releases (Gopinath et al. 2020). This leads to widespread human and environmental exposure. There is strong epidemiological evidence linking certain plastic chemicals to reproductive, neurodevelopmental, immune, and metabolic disorders (Monclús et al. 2025).

6.4.2 CHANGES IN THE CONCENTRATION OF SUBSTANCES AT DIFFERENT TROPHIC LEVELS

Biomagnification (aka 'bioamplification' or 'biological magnification') is the increase in concentration of a substance in the tissues of organisms at successively higher trophic levels in a food chain (see Chapter 5) (Silvy 2012). Two related terms are 'bioconcentration' and 'bioaccumulation'. Bioconcentration occurs when the uptake of a substance from water is greater than that which is excreted and bioaccumulation when the concentration of a substance in certain tissues increases due to absorption from food and the environment (Landrum and Fisher 1999). In other words, biomagnification occurs across trophic levels while bioconcentration and bioaccumulation occur within an organism. There is also 'biodilution' which occurs in aquatic ecosystems when the concentration of substance decreases in each higher trophic level.

Many chemicals that bioaccumulate are highly soluble in fats (lipophilic) and insoluble in water (hydrophobic). Global attention was drawn to the issues of biomagnification by Rachel Carson (1907–1964, an American biologist, writer, and conservationist) in her 1962 bestseller, *Silent Spring*. The classic example of

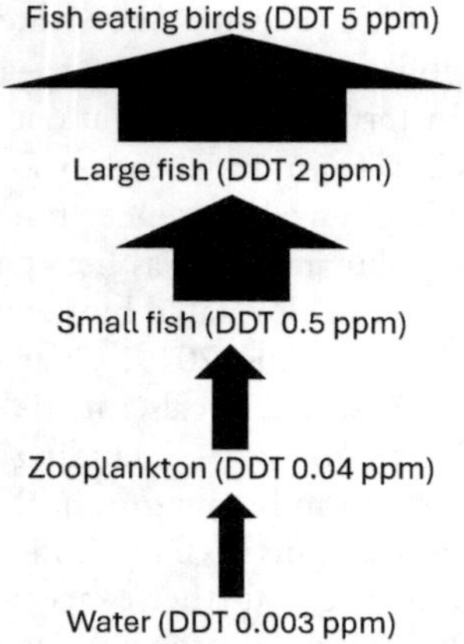

Figure 6.8 Biomagnification of DDT (dichlorodiphenyltrichloroethane) across trophic levels.

biomagnification is DDT (dichlorodiphenyltrichloroethane). DDT was first synthesised in 1874. It was widely used to combat diseases such as malaria (spread by female *Anopheles* mosquitoes) and typhus (spread by lice, mites, and fleas – all invertebrates) before being banned in the late 1970s. DDT is a persistent organic pollutant (POP). It accumulates progressively in the adipose tissue of the animals and at each trophic level its concentration is magnified (Figure 6.8). Such magnification has lethal and non-lethal effects such as eggshell thinning in raptors resulting in a decline in the number of breeding successes and the population size. Other substances that biomagnify are polychlorinated biphenyls (PCBs), which were used in coolant liquids, heavy metals, such as mercury, lead, and cadmium, and micro- and nanoplastics.

6.5 NON-EQUILIBRIUM ECOLOGY

6.5.1 NON-EQUILIBRIUM SYSTEMS

The discussion in Chapter 5 followed the idea that communities are conceptualised as stable and deterministic following a trajectory of community succession moving from one stage to another with set and predictable changes in the diversity and abundance of species present and associated abiotic factors (e.g. soil fertility). These ideas represent an equilibrium view of ecosystems. Non-equilibrium ecology asserts that ecosystems are dynamic and stochastic (i.e. may not be predicted precisely). From a non-equilibrium viewpoint, ecosystems have a limited capacity for change and reach thresholds at which the system shifts to an altered state (often relatively quickly). Such a shift can be seen in changes in the species composition and dominance, and in ecosystem processes. Briske (2017) provided a concise table comparing non-equilibrium and equilibrium systems (Table 6.2).

Non-equilibrium systems, rather than moving along a single continuum of change, flip between two states. Flipping back from a switched state requires significant change in the system. The example illustrated in Figure 6.9 is of a freshwater lake, which exists in either an oligotrophic or eutrophic state. In the oligotrophic state the lake has low nutrient levels, contains low to moderate levels of plant production (i.e. has a relatively low population of algae and plants), and has relatively clear water. As nutrient levels in an oligotrophic lake build up, the lake flips to the eutrophic state. In the eutrophic state plant net primary productivity is high, the water is murky and may be anoxic, and there may be blooms of toxic blue-green alga (Cyanobacteria). In this eutrophic state, fish and waterfowl in the lake stir up sediment and so produce a positive feedback loop that recycles nutrients (particularly phosphorus) and

Table 6.2 Comparison of non-equilibrium and equilibrium systems

Type	*Non-equilibrium systems*	*Equilibrium systems*
Abiotic patterns	Stochastic (i.e. may not be predicted precisely)/variable	Relatively constant
Plant-herbivore interactions	Weakly coupled Driven by abiotic factors	Tightly coupled Driven by biotic factors
Population patterns	Density independent Populations vary as carrying capacity changes	Density dependent Populations track carrying capacity
Community/ ecosystem characteristics	Structure driven by external factors	Structure driven by internal factors (competition)

Source: based on Briske (2017)

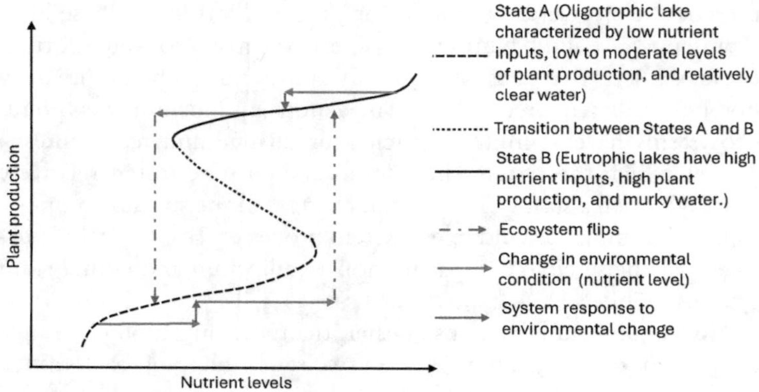

Figure 6.9 Non-equilibrium flipping of ecosystem between two states in a hypothetical lake.

maintains the eutrophic state. If there are substantial populations of algal predators and macrophytes in the lake, the feedback loops are weakened. The lake sediment becomes stabilised which effectively removes phosphorus from the lake. If sufficient phosphorus is removed, the lake will flip back to its oligotrophic state (Peterson 2008).

6.5.2 ECOSYSTEM STABILITY AND RESILIENCE

Ecosystems are subject to both shocks and stresses. Shocks are short duration events (e.g. a short-lived flood), which affect one or more ecosystem variable. Stresses affect ecosystem variables for a longer time (e.g. persistent poaching or the chronic presence of a pollutant). When viewed as a dynamic system, ecosystems are expected to change over time, but the question is, how do they respond to shocks and stresses and at what point do they cease to function. In other words: how stable is any given ecosystem?

Stability may be defined either as the difficulty of deflecting an ecosystem from its current composition and/or functioning, or in terms of how rapidly it returns to its original composition and functioning following a disturbance (Figure 6.10).

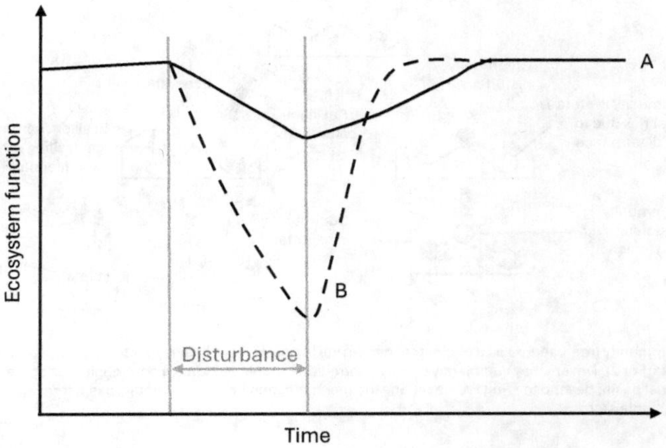

Figure 6.10 Two possible responses of an ecosystem to a disturbance. (A) Inertia – little change in function over time. (B) Resilience – initial large change function and then return to pre-disturbance function.

Source: redrawn with amendments from Leps (2013)

In the case of the first definition, inertia (i.e. resistance to change) is the key property of the system. In general terms an ecosystem with a high inertia shows little change in the face of a disturbance. In the second definition of stability – how quickly an ecosystem returns to its former state – it is resilience that is the key property. Resilience is the capacity of an ecosystem to tolerate disturbance without flipping to a qualitatively different state that is controlled by a different set of processes (*sensu* Holling 1973) (Figure 6.11). A more resilient ecosystem returns to its original state more rapidly. Ecosystems with a high degree of resiliency are dominated by negative feedback loops, while ecosystems with a low degree of resiliency are dominated by positive feedback loops. Feedback loops are ecosystem processes that influence variable rates of change. Negative feedback loops have a stabilising effect while positive feedback loops amplify and accelerate change.

Examples of key resilience ecosystem processes are shown in Figure 6.12. Some of these processes (e.g. diversity, heterogeneity,

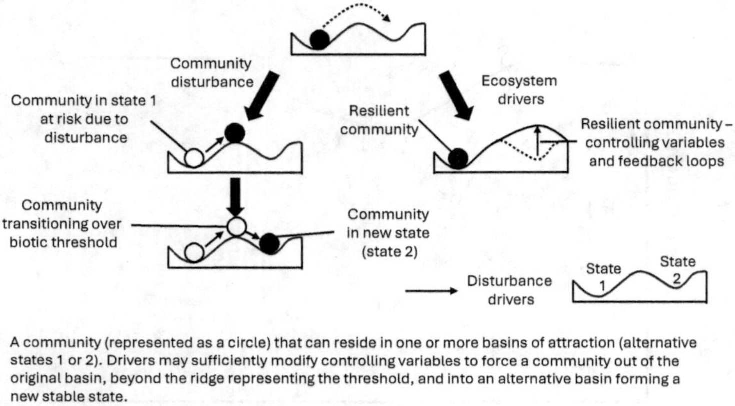

Figure 6.11 The basin of attraction model.
Source: adapted from Briske (2017)

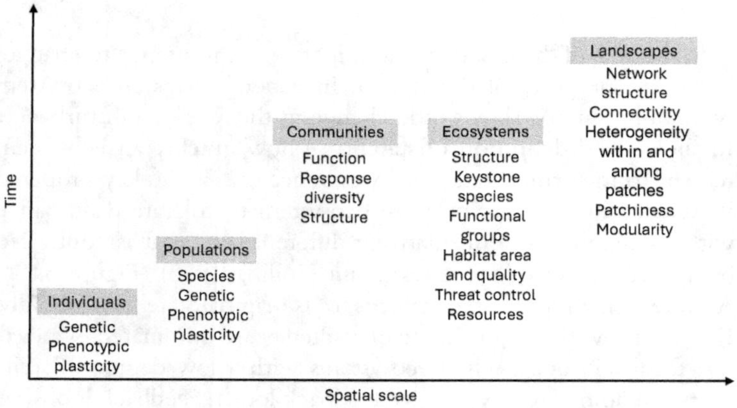

Figure 6.12 Conceptual diagram of key ecological resilience ecosystem processes relevant to individuals, populations, communities, ecosystems, and landscapes (i.e. organisational units).
Source: modified from Standish and Parkhurst (2024)

and connectivity) are important considerations within each spatial level. Resilience is maintained by the responses (and effects) of species in ecosystems and across landscapes (Mori et al. 2013), that is to say that it is 'response diversity' that is considered critical for ecological resilience of ecosystems to disturbance. Response diversity comes from both genetic diversity and phenotypic plasticity. Genetic diversity within populations, or within a species across its range (i.e. adaptive variation), enables organisms to respond to environmental change (Breed et al. 2013). Phenotypic plasticity, which is an organism's ability to adjust their physiology, morphology, and/or behaviour in response to an environmental challenge, can be observed in a broad range of species and traits (Nicotra et al. 2010). Response diversity contributes to functional compensation (i.e. extant species compensate functionally for species that are lost so that overall functioning is maintained) and to niche complementarity (e.g. differences in resource use between species) (Mori et al. 2013).

Knowledge and understanding of these resilience mechanisms facilitates the design and implementation of interventions to maintain and recover these mechanisms (Table 6.3). The examples presented in Table 6.3 are not exhaustive, but they do give an indication of approaches that are available and how they link back to resilience mechanisms.

There are some issues with the concept of resilience when applied to ecosystems. For example, simpler ecosystems may appear to be more resilient than more complex ecosystems – a neglected arable field covered with early succession weeds may be disturbed and yet return to its original state very rapidly, whereas a patch of forest, with late succession plants, may take much longer to return to its original state following a disturbance (i.e. be less resilient). This can be misleading as too tight a focus on resilience may not yield the desired conservation goal.

A different approach is to combine both the ideas of inertia and resilience into predictability. This allows for a stable ecosystem to be thought of as one that behaves in a predictable manner independent of what shocks or stresses it faces. This is why many conservation programmes strive to establish a kind of 'biological insurance' against situations in which stresses negatively affect sensitive species by maintaining or creating biodiversity rich ecosystems. Biological

Table 6.3 Interventions to maintain and restore resilience mechanisms

Organisational unit (level)	*Resilience mechanisms*	*Interventions*
Individuals and populations	Genetic diversity Phenotypic plasticity	Implement assisted migration (e.g. translocate plants and animals)
Community	Habitat heterogeneity Response diversity Structure	Manage and restore for native species diversity Consider response diversity and resource use complementarity Add artificial structures or materials to overcome time-lags the development of habitat structure (e.g. install hollow logs, nest boxes, woody debris)
Ecosystem	Keystone species and functional groups Habitat area Habitat quality Threat control	Reduce barriers to establishment of keystone species and functional groups Control of invasive species Reduce resources for introduced fauna species (e.g. fencing artificial water points to prevent access to species which are not desired in the ecosystem)
Landscape	Ecosystem network structure Connectivity Heterogeneity within and among patches Modularity	Restore degraded land Increase landscape-scale native vegetation cover, patch connectivity, and diversity

Source: based on Standish and Parkhurst (2024)

insurance theory suggests that in environments subject to temporal and spatial fluctuations, ecosystems with higher species diversity exhibit greater stability. A more diverse community is more stable and resilient to environmental fluctuations or disturbances because different species respond differently to changes. The portfolio effect is a core mechanism of biological insurance theory and states that the temporal variation of an aggregate ecosystem property, like total community biomass, will be less than the variation of its individual components (species). This is because species with asynchronous population dynamics can compensate for one another.

In other words, when the abundance of some species is negatively affected by a disturbance, the abundance of species that are more tolerant or even benefit from the change can increase, ensuring that overall ecosystem function and services remain stable. Many conservationists follow these ideas and work towards maintaining and enriching biodiversity in ecosystems rather than being focussed on the individual species within that ecosystem. This will be a theme picked up in Chapter 7.

6.6 NOVEL ECOSYSTEMS (AKA EMERGING ECOSYSTEMS OR NO-ANALOGUE ECOSYSTEMS)

Evidence that drivers including land transformation and climate change associated with human activities are permanently changing ecosystems began to emerge in the latter part of the twentieth century. Chapin and Starfield (1997), while reporting on modelling work on Arctic Alaska, are considered the first to mention the term 'novel ecosystem' in peer reviewed literature. In the early 2000s this term and 'emerging ecosystems' (which is often used as a synonym) have been used more frequently. Examples of novel ecosystems can be found in both aquatic and terrestrial environments and include algal-dominated coral reefs in polluted warming seas, formerly cultivated land that is now dominated by non-native species and an altered fire regime (i.e. the patterns, frequency, and intensity of bushfires and wildfires in an area over long periods), and woodlands with changed sub-canopy communities (Kujper et al. 2024).

Hobbs et al. (2013) defined a novel ecosystem as a system of abiotic, biotic, and social components (and their interactions) that differ from those that prevailed historically and which self-organise and manifest novel qualities without intensive human management. Novel ecosystems typically comprise new species assemblages, have interactions and processes that are consistent with a self-organised, functioning ecosystem, and have considerable/significant biotic and/or abiotic alterations that make them difficult to 'restore' to a previous condition (Harris et al. 2024).

A key aspect of novel ecosystems is that they can include species that which are outside the 'historical' native ranges. This can be as a result of anthropogenic introduction (both intended and unintended) or through 'natural' range expansion (Kujper et al. 2024).

Tropical species are increasingly expanding their range away from the equator as a result of climate change and the associated expansion of the tropical climate zone. At the same time, land use and land cover changes driven by conversion of land to agriculture and suburban environments is resulting in the expansion of disturbance-tolerant species (Stokes and Keitt 2025).

Hybridisation between species that had not previously occurred in the same geographic location is one potential outcome of novel ecosystems. Stokes and Keitt (2025) reported examples of hybridisation occurring driven by human-assisted species introductions and habitat modification via disturbance or altered fire regimes. They also reported a natural hybridisation, the result of a female green jay (*Cyanocorax yncas*) breeding with male blue jay (*Cyanocitta cristata*), due to climate change and urbanisation. The green jay is a tropical species which is expanding its range from the northern Andes through Central America and Mexico into southern Texas by as much as 2 degrees of latitude over the span of a few generations. The blue jay has expanded its range to south and west Texas during a similar time period apparently tracking the expansion of human settlement. The range of the two species now overlap.

From a restoration or conservation viewpoint, novel ecosystems may be viewed as being unwelcome as they cannot or cannot easily be removed to establish historical ecosystems. However, novel ecosystems can benefit endangered plants and animals that are threatened by habitat loss elsewhere, and they also provide a valuable connection for humans with nature in peri-urban environments (Harris et al. 2024).

6.7 SOCIAL-ECOLOGICAL SYSTEMS

Social-ecological systems are based on the concept that humans are a part of nature and, hence, part of the same system. This idea was first put forward in 1998 by Berkes (1945–, a Turkish Canadian ecologist) and Folke (1955–, a Swedish trans-disciplinary environmental scientist) (Berkes and Folke 1998), and further developed by Berkes et al. (2003) and Winter et al. (2018). In developing their ideas these authors recognised that the diversity of life comprises plants and animal species, habitats, ecosystems, and biocultural diversity (i.e. the diversity of human cultures and languages). They also recognised that biocultural diversity is critical to the structure, function, and resilience of social-ecological systems (Winter et al. 2018).

A social-ecological system consists of 'a bio-geo-physical' unit, such as a woodland, and the associated social actors and institutions (e.g. landowners and managers, conservation organisations, individuals and social groups, local authorities, and national government and associated governmental departments). That is to say that a social-ecological system combines natural factors (e.g. biodiversity, land use and land cover, soil, nutrient fluxes, water (fresh, estuarine, and marine), air (including stratospheric ozone and aerosols), diseases (of plants, animals, and humans), climate and weather, and contaminants in soil, water and air with social factors (e.g. human demographics, cultural, social, technological, regulatory, economic, and financial) (Tzoulas et al. 2021).

Social-ecological systems are delimited by spatial or functional boundaries surrounding particular ecosystems and their contextual problems. They are complex (i.e. they are composed of many components which may interact with each other) and adaptive (i.e. they are hierarchical systems that responds to environmental changes or changes in the interacting social or ecological parts, generally through feedback loops) (Figure 6.13).

Systems such as that outlined in Figure 6.11 are a set of interacting or interdependent entities that form an integrated whole and together can respond to environmental changes or changes in the interacting parts (i.e. they are adaptive systems). Adaptive cycles have certain properties which are outlined in Table 6.4.

An adaptive cycle passes through four recognised phases: *r* (exploitation), *K* (conservation), Ω (omega – release), and α (alpha – reorganisation) (Figure 6.14). In this nomenclature *r* and *K* relate to the *r* and *K* life strategies (see Chapter 4, section 4.5) where *r* indicates a stage where the environment has been disrupted in some way and is favourable to *r*-selected species, and *K* to a stage when the environment is more or less stable and is favourable to *K*-selected species can survive at or near their carrying capacity. Movement through each phase of the adaptive cycle is dependent on the interaction of social and ecological systems. For example, if a mature woodland is cleared for timber, that parcel of land has moved from the *K* (conservation) phase to the Ω (release) phase. Clear felling a woodland is a societal choice determined by factors

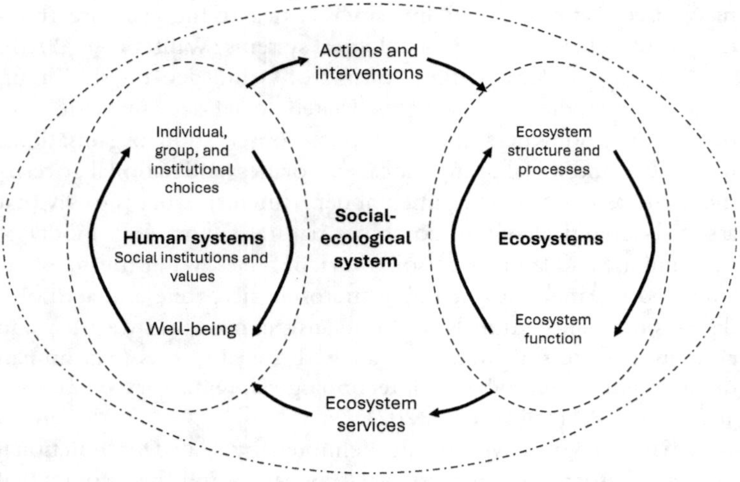

Figure 6.13 Simplified schematic of a social-ecological system.

Table 6.4 Properties of an adaptive social-ecological system

Property	*Description*
System potential	There is a range of options available for the future behaviour of the system (e.g. accumulation of nutrients, resources, biomass, and diversity) in response to change.
Connectedness	The relationships between system elements and processes, and the degree to which elements are dominated by external variability, or by relationships that mediate the influence of external variability (Holling and Gunderson 2002).
Resilience	The degree of disturbance a system can buffer without moving into a new regime, or basin of attraction (i.e. a set of states toward which a system tends to evolve, for example, a wet land moving towards grassland due to reduced rainfall) (Holling 1973).

such as economics and the demand for timber. Movement from the Ω (release) phase to the α (reorganisation) phase will be dependent on the interaction of choices made about the future of that parcel of land and the ecological factors. Examples of choices include to restock, change the tree species present, or rewild the area. Similarly,

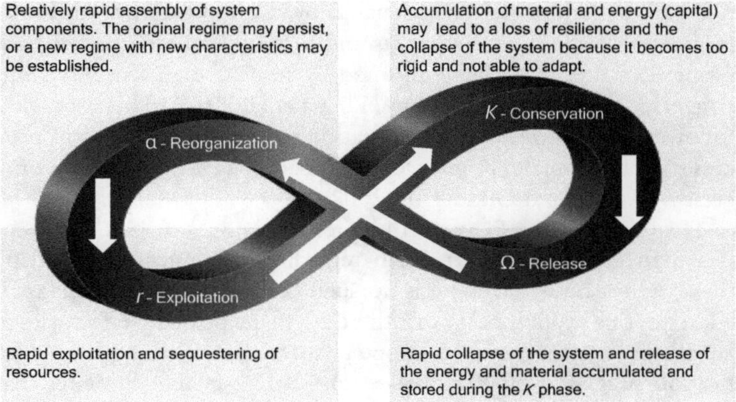

Figure 6.14 Two-dimensional representation of the adaptive cycle as originally conceived by Gunderson and Holling (2002).

movement from the α (reorganisation) phase to the *r* (exploitation) phase is dependent on the interaction of social and ecological factors which determine how habitats on that land develop ecologically. Finally, there will be an interaction of ecological and societal factors such as the occurrence of disturbances (e.g. fire, land management actions, the occurrence of pests and diseases, and social factors related to the use of the land).

Figure 6.13 illustrates how an adaptive cycle system moves through a 3-dimensional state in which space is defined by system potential, connectedness, and resilience (Holling 1986; Gunderson and Holling 2002). Movement through the phases is not rigid or predetermined. In the *r* and *K* phases, there is an accumulation of resources and relationships: initially relatively rapid in the *r* phase but slower as the *K* phase is reached. As the system moves from the *r* phase into the *K* phase, potential and connectedness increase, but resilience shrinks due to the high connectedness amongst system elements making the system vulnerable to cascading disturbances. Systems tend to spend most time in these phases, and the *K* phase is often understood to be relatively stable over time with regards to properties such as species richness, diversity, and biomass (Hatton et al. 2015; Allhoff and Drossel 2016). In the two back loop phases (Ω and α) the system transitions rapidly to a new *r* (exploitation) phase. The Ω phase is characterised by the rapid release of accumulated resources (e.g. nutrients and biomass) when a disturbance event (e.g. fire, drought, consumed by insects, or intense grazing) triggers a collapse of the system (Thapa et al. 2016). This is quickly followed by a period of reorganisation (α), such as when soil processes capture nutrients and pioneer species begin colonisation processes (Holling and Gunderson 2002).

The way in which Figure 6.14 is presented might suggest that there is a high degree of determinism regarding the trajectory of adaptive systems: however, there is no such certainty. A second diagram addresses this by illustrating that there are many possible trajectories as an adaptive system moves from one phase to the next. This is a modification suggested by Burkhard et al. (2011). It also illustrates that there is no increase in resilience at the end of the Ω (release) phase as the system can move in any one of many directions (Figure 6.15).

Ecosystems contain scale domains (also referred to as spatial and temporal domains, e.g. a tree, a woodland, and a wooded landscape) that are separated by discontinuities, or scale breaks that represent a transition to a new set of structuring processes (i.e. the processes operating at the scale of an individual tree are not the same as those

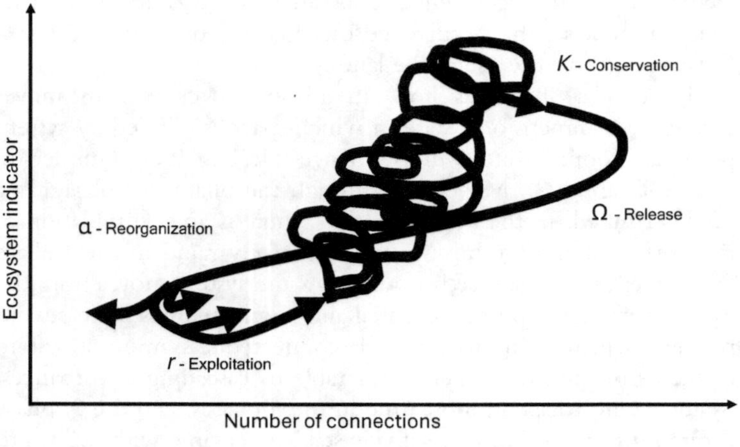

Figure 6.15 More realistic depiction of the variability within an adaptive social-ecological cycle.

Source: based on Burkhard et al. (2011)

operating at the scale of a woodland). Thus, adaptive cycles occur at each scale domain within the system. At a global scale, ecosystems (known as biomes at this scale) relevant structuring processes include latitude and climate zones, while at the regional scale factors such as land use, water management, and human activities become important. At the landscape scale topography, vegetation, and soil type are important factors, and at the scale of an individual woodland or grassland management practices, species richness and abundance become important. This results in complex systems with multiple and nested domains of scales at different phases of the adaptive cycle. There is some interaction between the scale domains with occasional changes being cascaded up and down between domains. Nested adaptive cycles with bi-directional cross-scale feedback loops are called a Panarchy (Figure 6.16).

6.8 ECOSYSTEM SERVICES AND BENEFITS, AND NATURAL CAPITAL

In the preceding sections of this chapter, the role of ecosystems in cycling nutrients and pollutants has been discussed. These are two of the services provided by ecosystems. The idea of 'ecosystem services' originated in the 1970s and has gained in popularity over time. Ecosystem services are essential for the maintenance of ecosystems, human well-being, and economic development. Various definitions of ecosystem services have been proposed, which all focus on the benefits provided by ecosystems.

The authors of the Millennium Ecosystem Assessment (2005) divided ecosystem services into four categories: supporting, provisioning, cultural, and regulating (Figure 6.17).

An example of how this classification can be applied to the ecosystem services provided by the sea is presented in Figure 6.18.

More examples of ecosystem services are presented in Table 6.5.

One ecosystem may provide several services. For example, marine ecosystems provide provisioning services including marine resources and genetic resources, supporting services including nutrient cycling and primary production, regulating services including carbon sequestration and flood control, and finally cultural services including recreation and tourism. Figure 6.19 illustrates the diverse ecosystem services that are associated with a mangrove swamp.

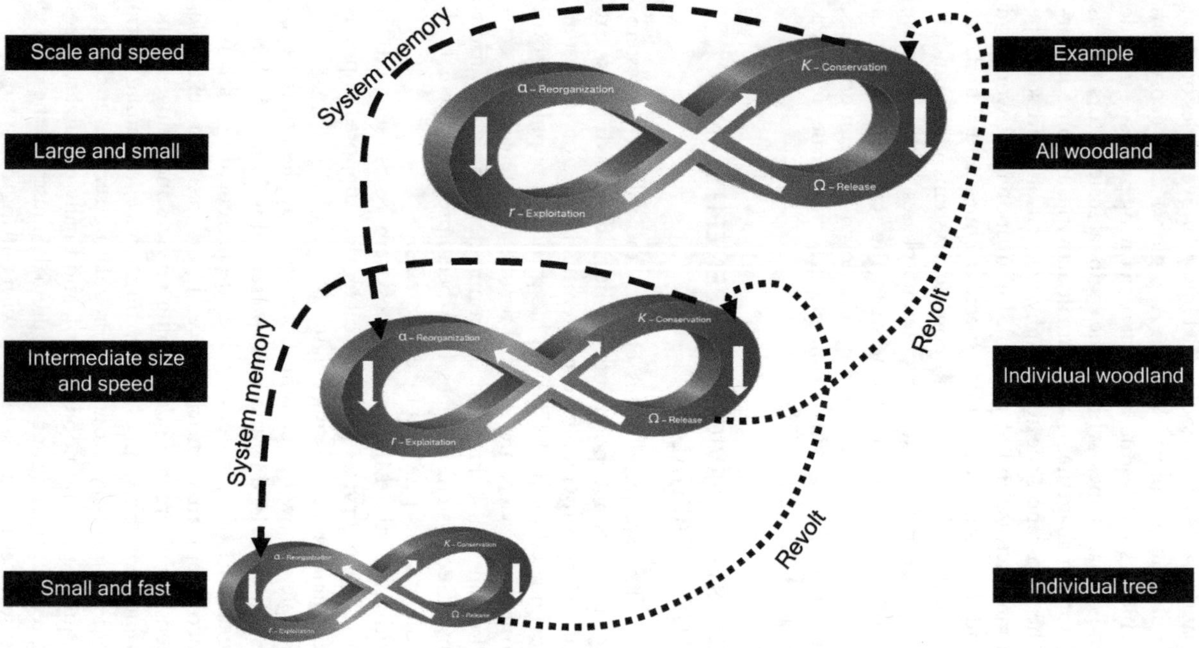

Figure 6.16 The classic Panarchy figure, showing three nested adaptive social-ecological cycles with bi-directional cross-scale feedback loops.

Source: based on Gunderson and Holling (2002)

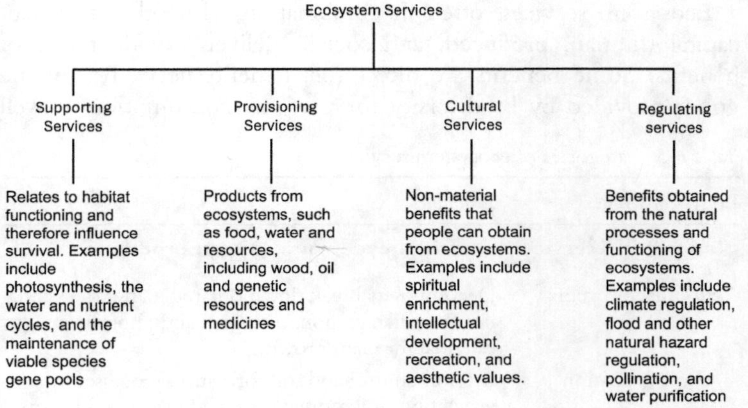

Figure 6.17 Classification of ecosystem services used in the Millennium Ecosystem Assessment (2005).

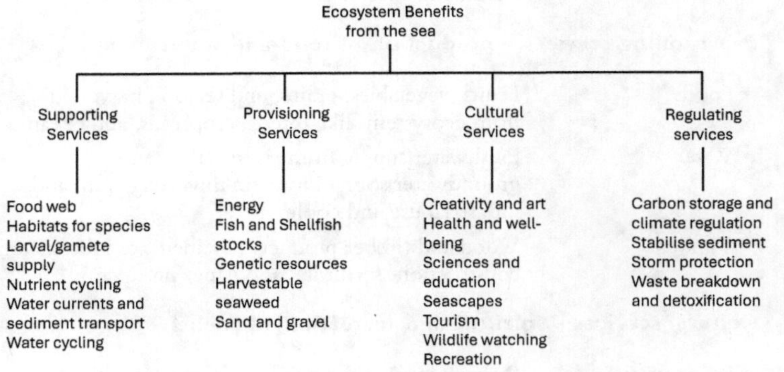

Figure 6.18 Ecosystem services associated with the sea.
Source: based on NatureScot (2020)

The concept of 'natural capital' recognises that certain elements of nature, both the abiotic and biotic, have a value to society. The amount of an element available is termed natural capital stocks (e.g. there are stocks of fish, timber, water, land, biodiversity, and minerals). Ecosystem services (sometimes called environmental services) flow over time from these stocks of natural capital.

Ecosystem services, often in combination with other forms of capital (human, produced, and social), deliver a wide range of benefits. Some benefits are direct (i.e. benefits derived from the goods provided by biodiversity for personal consumption as well

Table 6.5 Categories of ecosystem services

Ecosystem services	*Examples*
Supporting services – nutrient cycles and oxygen production	
• Nutrient cycling	Decomposers break down organic matter, recycling nutrients like carbon, nitrogen, and phosphorus essential for plant growth.
• Soil formation	Erosion control and soil formation processes by vegetation, microorganisms, and geological factors sustain fertile soils for agriculture and ecosystem health.
• Biodiversity	Rich biodiversity supports ecosystem resilience, genetic diversity, and evolutionary processes, enhancing ecosystem stability and productivity.
Provisioning services – production of food and water	
• Food	Fruits, vegetables, grains, and seafood harvested from ecosystems like forests, croplands, and oceans.
• Water	Freshwater supply from rivers, lakes, and groundwater sources for drinking, irrigation, and industrial use and cooling.
• Timber	Wood and timber products obtained from forests for construction, furniture, and paper production.
Cultural services – spiritual and recreational benefits	
• Recreation	Parks, forests, and natural landscapes provide opportunities for outdoor activities like hiking, camping, and wildlife observation, promoting physical and mental well-being.
• Spiritual enrichment	Natural environments offer spiritual and aesthetic experiences, fostering a sense of connection to nature, tranquillity, and inspiration.
• Cultural heritage	Indigenous knowledge, traditions, and practices associated with ecosystems preserve cultural identity, heritage sites, and traditional ecological knowledge.

(*Continued*)

Table 6.5 (Continued)

Ecosystem services	*Examples*
Regulating services (aka maintaining services) – control of climate and disease	
• Climate regulation	Forests and oceans absorb carbon dioxide, helping mitigate climate change by regulating the Earth's temperature.
• Water purification	Wetlands and riparian zones filter pollutants and purify water, improving water quality for aquatic ecosystems and human consumption.
• Pollination	Bees, butterflies, and other pollinators facilitate agricultural productivity and biodiversity.

Source: based on the Millennium Ecosystem Assessment (2005)

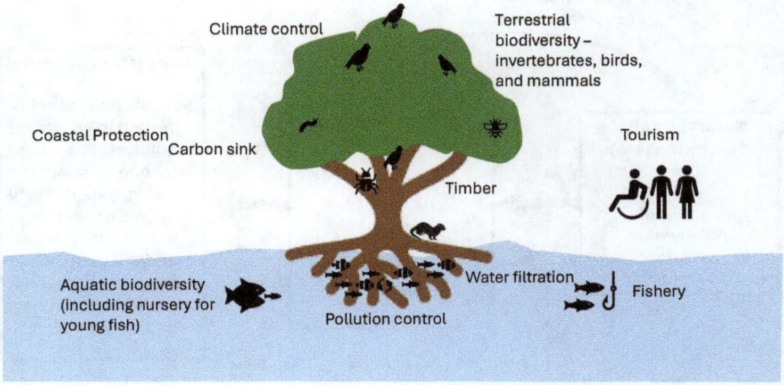

Figure 6.19 Ecosystem services provided by a mangrove swamp.

as generating income through trade), and others are indirect (i.e. benefit that is of great value to society, not individuals or businesses) (Table 6.6).

Each of these benefits has a value. Some benefits relate to the use to which the resource is put. Such benefits have a market value (e.g. minerals, timber, and freshwater). Other uses are recognised for their non-market value. This is the non-market value people place on the existence of habitats or species (e.g. outdoor recreation and landscape amenity) (Figure 6.20).

Table 6.6 Example of direct and indirect ecosystem benefits

Direct ecosystem benefits (benefits for individual consumption and trade)	*Indirect ecosystem benefits (benefits for society rather than for individuals or businesses)*
Food Timber Fuel (e.g. firewood, oil, gas) Medicines Fibre (linen and wool), and Freshwater	Climate regulation through carbon sequestration Water purification Flood control Decomposition of dead matter by bacteria and fungi Pollination of crops Nutrient cycling that maintains soil fertility and ecosystem productivity Landscapes for recreation Sacred forests

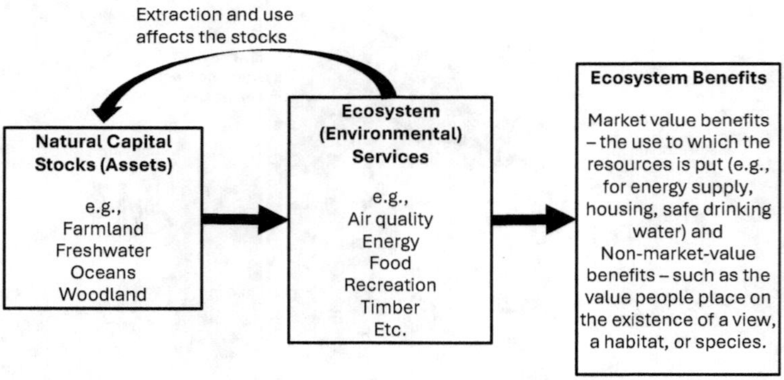

Figure 6.20 Relationship between natural capital stocks, ecosystem services, and ecosystem benefits.

The quality, quantity, and location of the natural stocks/assets determine how easily the services and benefits may flow and be realised. Background pressures, management practices, and drivers of demand also affect this system. To recognise fully the benefits from some services, additional economic and processing inputs are required. In other cases, the benefit follows directly from the stock, through the service, to the benefit without further capital or human input.

6.9 CASE STUDY – NATURE-BASED SOLUTIONS

The nature-based solutions (NbS) approach emerged during the late 2000s. The essence of the concept is to work with ecosystems to adapt to and mitigate climate change effects (see Chapter 2), improve sustainable livelihoods, and protect natural ecosystems and biodiversity. The IUCN defines NbS as:

> actions to protect, conserve, restore, sustainably use and manage natural or modified terrestrial, freshwater, coastal, and marine ecosystems, which address social economic, and environmental challenges effectively and adaptively, while simultaneously providing human well-being, ecosystem services and resilience and biodiversity benefits.
>
> (Cohen-Shacham et al. 2016, p. xii)

The introduction of the NbS approach marked a subtle shift in perspective: people were not just the passive beneficiaries of nature, they could also purposefully and proactively protect, manage, or restore natural ecosystems to make a significant contribution to addressing major societal challenges. With this shift of perspective comes the concept that people have a responsibility to manage the environment for their benefit and that of future generations. There was a second important shift in perspective: a broadening of the scope of nature conservation from an exclusive focus on the protection of wilderness and wild, charismatic species to approaches that tackle the drivers of biodiversity decline such as pollution and the loss of habitats. These two shifts in perspective mean that the conservation agenda now embraces a more complex understanding of social-ecological systems. This requires an interdisciplinary approach (see Chapter 1), an understanding of communities (see Chapter 6), ecosystems (this chapter), and social-ecological systems (this chapter).

NbS interventions can take many forms (Table 6.7). NbS can be standalone or employed in conjunction with other types of interventions. Nature-based solutions are a departure from engineering-based approaches to addressing environmental issues. For example, flood alleviation is often associated with civil engineering works to instal flood barriers, canalise rivers to increase flow rates, and

Table 6.7 Examples of nature-based solutions (NbS)

NbS	*Nature benefits*	*Societal and economic benefits*
Restore and sustainably manage wetlands and rivers	Maintain or boost fish stocks	Maintain or boost fisheries-based livelihoods Reduce the risk of flooding, and provide recreational Benefit tourism
Conserve forests	Support biodiversity	Support food and energy security, local incomes, and climate change adaption and mitigation
Restore drylands	Support biodiversity	Strengthen water security, local livelihoods and resilience to climate change impacts
Developing green infrastructure in urban environments (e.g. green walls, roof gardens, street trees, vegetated drainage basins)	Support biodiversity	Improve air quality, support wastewater treatment Reduce stormwater runoff and water pollution Improve the quality of life for residents
Use natural coastal infrastructure (e.g. barrier islands, mangrove forests, and oyster reefs)	Support biodiversity	Protect shorelines and communities from coastal flooding Reduce the impacts of sea-level rise

Source: based on examples in Cohen-Shacham et al. (2016)

construct flood attenuation basins. A nature-based solution approach would see interventions that facilitate the holding of water in river catchments to reduce the amount of water entering the river and, hence, attenuate downstream flooding. Examples of such NbS include promoting water infiltration and the (re-)installation of meandering rivers with associated flood plains. Such thinking requires support from local communities and stakeholders right from the planning stage through to implementation to address concerns around the approach.

A typology of NbS was proposed by Eggermont et al. (2015) (Table 6.8). This typology is helpful as a means of framing the NbS concept and considering the kinds of approaches that would classify as NbS.

Table 6.8 A typology of nature-based solutions (NbS)

NbS type	*Description*	*Examples*	*Ecosystem services delivery*	*Level of engineering applied to biodiversity and ecosystems*
Type 1	Use existing natural or protected ecosystems	Measures to increase fish stocks in an intact wetland to enhance food security	Low	Low
Type 2	Manage or restore ecosystems	Re-establishing traditional agro-forestry systems based on commercial tree species to support poverty alleviation	Moderate	Moderate
Type 3	Create new ecosystems	Establishing green buildings (e.g. green walls, green roofs)	High	High

Source: based on Eggermont et al. (2015)

Table 6.9 Examples of nature-based solution (NbS) approaches

NbS approach	*Examples*
Ecosystem restoration approaches	Ecological restoration Ecological engineering Forest landscape restoration
Issue-specific ecosystem-related approaches	Ecosystem-based adaptation Ecosystem-based mitigation Climate adaptation services Ecosystem-based disaster risk reduction
Infrastructure-related approaches	Natural infrastructure Green infrastructure
Ecosystem-based management approaches	Integrated coastal zone management Integrated water resources management
Ecosystem protection approaches	Area-based conservation approaches including protected area management

Within the umbrella concept of NbS there is a range of ecosystem-related approaches all of which address societal challenges (Table 6.9).

In practical terms the need to incorporate a diversity of responses into a nature-based solution (see section 6.5.2) can be addressed by deliberately incorporating or maintaining species with different response (and effect) traits (Aquilué et al. 2020; Walker et al. 2023). The goal is to create and maintain diverse stands as these are more likely to contain the variety of species' responses needed to maintain ecosystem function in the face of disturbance. In the NbS approach species can be both functionally unique as well as redundant (Mori et al. 2023; Zavaleta et al. 2010). This is sometimes referred to as a 'portfolio effect' under the biological insurance theory (see section 6.5.2) (*sensu* Yachi and Loreau 1999; Loreau et al. 2021). This approach becomes more important when more than one function is desired, which applies to majority of nature-based solutions.

Okello et al. (2024) reviewed critical NbS interventions, including water conservation, soil moisture and conservation, water harvesting, conservation agriculture, agroforestry, and afforestation in arid and semi-arid lands in Africa. The potential of NbS to improve water availability, ecosystem resilience, and socio-economic development were highlighted in the review. Barriers to implementation were found to include limited stakeholder involvement, inadequate integration of indigenous knowledge, and regional research disparities. This emphasises the importance of stakeholder engagement at all stages of ecological based interventions. This is a theme that will be picked up in the Chapter 7.

6.10 SELF-STUDY SUGGESTIONS

6.10.1 SELF-STUDY QUESTIONS

1. What is an ecosystem?
2. What are the major components of an ecosystem and how are they related?
3. How do energy and materials flow within and through an ecosystem?
4. What is the meaning of primary production, gross primary productivity, and net primary productivity?
5. What are the problems associated with the concept of trophic levels?

6. What is a trophic cascade?
7. What is a pyramid of energy and how does it compare with a pyramid of numbers and a pyramid of biomass (see Chapter 5)?
8. What are the similarities and differences between the oxygen and carbon cycles and the plastics cycle?
9. What roles are played by microorganisms in the nitrogen cycle?
10. What is the role of carbon and nitrogen in decomposition?
11. What is biomagnification and what are the consequences of this process?
12. What are ecosystem services and how do they relate to ecosystem benefits?
13. What the characteristics of a non-equilibrium system?
14. Why do non-equilibrium systems flip rather than continue along a continuum and what are the implications of this?
15. What is a social-ecological system?
16. What are the properties of an adaptive system?
17. What are the consequences of adaptive systems for conservation?

6.10.2 A DEEPER DIVE

A. Consider your diet for one week and workout your food web. What tropic level(s) do you occupy?
B. A useful resource is https://byjus.com/biology/ecosystem.
C. Conservation Evidence (www.conservationevidence.com) is a free, authoritative information resource that supports decisions about how to maintain and restore global biodiversity. On that site there are summaries and synopses of evidence from the scientific literature about the effects of conservation actions such as methods of habitat or species management. New evidence is also published online in *Conservation Evidence Journal*.
D. Consider reading Colinvaux (1978).
E. Consider reading Kimmerer (2013).
F. Important biogeochemical cycles include carbon, oxygen, sulphur, nitrogen, phosphorus, water, heavy metals, plastics, pesticides, radioactive isotopes, nutrients, and several toxic substances. Explore details of these different cycles and consider the role of humans in changing them.

6.11 REFERENCES

Allhoff, K.T. and Drossel, B. (2016). 'Biodiversity and ecosystem functioning in evolving food webs'. *Philosophical Transactions of the Royal Society B* 371: 20150281. https://doi.org/10.1098/rstb.2015.0281.

Andrzejewska, L. and Gyllenberg, G. (1980). 'Small herbivore subsystem'. In A.I. Bretmeyer and G.M. Van Dyne (eds), *Grasslands, Systems Analysis and Man*, 201–268. Cambridge: Cambridge University Press.

Aquilué, N., Filotas, E., Craven, D., Fortin, M.-J., Brotons, L. and Messier, C. (2020). 'Evaluating forest resilience to global threats using functional response traits and network properties'. *Ecological Applications* 30(5): e02095. https://doi.org/10.1002/eap.2095.

Berkes, F., Colding, J. and Folke, C. (2003). *Navigating Social-Ecological Systems: Building Resilience for Complexity and Change*. Cambridge: Cambridge University Press.

Berkes, F. and Folke, C. (1998). *Linking Social and Ecological Systems: Management Practices and Social Mechanisms for Building Resilience*. Cambridge: Cambridge University Press.

Breed, M.F., Stead, M.G., Ottewell, K.M., Gardner, M.G. and Lowe, A.J. (2013). 'Which provenance and where? Seed sourcing strategies for revegetation in a changing environment'. *Conservation Genetics* 14: 1–10. https://doi.org/10.1007/s10592-012-0425-z.

Briske D.D. (ed). (2017). *Rangeland Systems: Processes, Management and Challenges*. Cham, Switzerland: Springer. https://doi.org/10.1007/978-3-319-46709-2.

Brylinsky, M. and Mann, K.H. (1973). 'An analysis of factors governing productivity in lakes and reservoirs'. *Limnology and Oceanography* 18: 1–14. https://doi.org/10.4319/lo.1973.18.1.0001.

Burkhard, B., Fath, B.D. and Müller, F. (2011). 'Adapting the adaptive cycle: hypotheses on the development of ecosystem properties and services'. *Ecological Modelling* 222(16): 2878–2890. https://doi.org/10.1016/j.ecolmodel.2011.05.016.

Carson, R. (1962). *Silent Spring*. Boston, MA: Houghton Mifflin Company.

Chamas, A., Moon, H., Zheng, J., Qiu, Y., Tabassum, T., Jang, J.H., Abu-Omar, M., Scott, S.L. and Suh, S. (2020). 'Degradation rates of plastics in the environment'. *ACS Sustainable Chemistry & Engineering* 8(9): 3494–3511. https://doi.org/10.1021/acssuschemeng.9b06635.

Chapin, F.S. and Starfield, A.M. (1997). 'Time lags and novel ecosystems in response to transient climatic change in Arctic Alaska'. *Climatic Change* 35: 449–461. https://doi.org/10.1023/A:1005337705025.

Coe, M.J., Cummings, D.H. and Philipson, J. (1976). 'Biomass and production in large African herbivores in relation to rainfall and primary production'. *Oecologia* 22: 341–354.

Cohen-Shacham, E., Walters, G., Janzen, C. and Maginnis, S. (eds.). (2016). *Nature-Based Solutions to Address Global Societal Challenges*. Gland: Switzerland: IUCN.

Colinvaux, P. (1978). *Why Big Fierce Animals Are Rare: An Ecologist's Perspective*. Princeton, NJ: Princeton University Press.

Dick, G.J. (2019). 'The microbiomes of deep-sea hydrothermal vents: distributed globally, shaped locally'. *Nature Reviews Microbiology* 17: 271–283. https://doi.org/10.1038/s41579-019-0160-2.

Eggermont, H., et al. (2015). 'Nature-based solutions: new influence for environmental management and research in Europe'. *GAIA* 24(4): 243–248. https://doi.org/10.14512/gaia.24.4.9.

Gaur, V.K., Sharma, P., Gaur, P., Varjani, S., Ngo, H.H., Guo, W., Chaturvedi, P and Singhania, R.R. (2021a). 'Sustainable mitigation of heavy metals from effluents: toxicity and fate with recent technological advancements'. *Bioengineered* 12: 7297–7313. https://doi.org/10.1080/21655979.2021.1978616.

Gaur, V.K., Sharma, P., Sirohi, R., Varjani, S., Taherzadeh, M.J., Chang J-S, Ng, H.Y.,Wong, J.W.C. and Kim, S-H. (2021b). 'Production of biosurfactants from agro-industrial waste and waste cooking oil in a circular bioeconomy: an overview'. *Bioresource Technology* 343: 126059. https://doi.org/10.1016/j.biortech.2021.126059.

Gopinath, K.P., Nagarajan, V.M., Krishnan, A. and Malolan, R. (2020). 'A critical review on energy, environmental and economic factors on various processes used to handle and recycle plastic wastes: development of a comprehensive index'. *Journal of Cleaner Production* 274: 123031. https://doi.org/10.1016/j.jclepro.2020.123031.

Gunderson, L.H. and Holling, C.S. (eds). (2002). *Panarchy: Understanding Transformations in Human and Natural Systems*. Washington, DC: Island Press.

Guo, X. and Wang, J. (2019). 'The chemical behaviors of microplastics in marine environment: a review'. *Marine Pollution Bulletin* 142: 1–14. https://doi.org/10.1016/j.marpolbul.2019.03.019.

Harris, J., Bullock, J., Pettorelli, N, Perring, M. and Mercer, T. (2024). '*What is a novel ecosystem?*'. *The Niche*, Autumn: 29–33.

Hatton, I.A., McCann, K.S., Fryxell, J.M., Davies, T.J., Smerlak, M., Sinclair, A.R.E. and Loreau, M. (2015). 'The predator–prey power law: biomass scaling across terrestrial and aquatic biomes'. *Science* 349: 6252. https://doi.org/10.1126/science.aac6284.

He, L., Chen, J.M., Pan, Y., Birdsey, R. and Kattge, J. (2012). 'Relationships between net primary productivity and forest stand age in US forests'. *Global Biogeochemical Cycles* 26(3): GB3009. https://doi.org/10.1029/2010GB003942.

Hobbs, R.J., Higgs, E. and Hall, C.M. (2013). 'Defining novel ecosystems'. In R.J. Hobbs, E. Higgs, and C.M. Hall, *Novel Ecosystems: Intervening in the New World Order*. Chichester: John Wiley & Sons. https://doi.org/10.1002/9781118354186.ch6.

Holling, C.S. (1973). 'Resilience and stability of ecological systems'. *Annual Review of Ecology, Evolution, and Systematics* 4: 1–23. www.jstor.org/stable/2096802.

Holling, C.S. (1986). 'The resilience of terrestrial ecosystems: local surprise and global change'. In W.C. Clark and R.E. Munn (eds), *Sustainable Development of the Biosphere*, 292–317. Cambridge: Cambridge University Press.

Holling, C.S. and Gunderson, L.H. (2002). 'Resilence and adaptive cycles'. In L.H. Gunderson and C.S. Holling (eds), *Panarchy: Understanding Transformations in Human and Natural Systems*, 25–62. Washington, DC: Island Press.

Hutchinson, G.E. (1970). 'The biosphere'. *Scientific American* 223(3): 44–53.

Kimmerer, R.W. (2013). *Braiding Sweetgrass*. Minneapolis, MN: Milkweed Editions.

Kuijper, D.P.J., Diserens, T.A., Say-Sallaz, E., Kasper, K., Szafrańska, P.A., Szewczyk, M., Stępniak,, K.M. and Chuski, M. (2024). 'Wolves recolonize novel ecosystems leading to novel interactions'. *Journal of Applied Ecology* 61: 906–921. https://doi.org/10.1111/1365-2664.14602.

Kutsch, W.L., Wirth, Christian, Kattge, Jens, Nöllert, Stefanie, Herbst, Matthias and Kappen, Ludger. (2009). 'Ecophysiological characteristics of mature trees and stands: consequences for old-growth forest productivity'. In C. Wirth, G. Gleixner, and M. Heimann (eds), *Old-Growth Forest: Function, Fate and Value*, 57–79. Berlin: Springer. https://doi.org/10.1007/978-3-540-92706-8.

Landrum, P.F. and Fisher, S.W. (1999). 'Influence of lipids on the bioaccumulation and trophic transfer of organic contaminants in aquatic organisms'. In M.T. Arts and B.C. Wainman (eds), *Lipids in Fresh Water Ecosystems*. New York: Springer.

Leps, J. (2013). 'Diversity and ecosystem function'. In E. Van der Maarel and J. Franklin (eds), *Vegetation Ecology*. Chichester: Wiley-Blackwell. pp 308–346.

Lomax, G.A., Powell, T.W.R., Lenton, T.M., Economou, T. and Cunliffe, A.M. (2024). 'Untangling the environmental drivers of gross primary productivity in African rangelands'. *Communications Earth and Environment* 5: 500. https://doi.org/10.1038/s43247-024-01664-5.

Loreau, M., et al. (2021). 'Biodiversity as insurance: from concept to measurement and application'. *Biological Review* 96: 2333–2354. https://doi.org/10.1111/brv.12756.

Millennium Ecosystem Assessment. (2005). *Ecosystems and Human Well-Being: Synthesis*. Washington, DC: Island Press.

Monclús, L., Arp, H.P.H., Groh, K.J., Faltynkova, A., Løseth, M.E., Muncke, J., Wang, Z., Wolf, R., Zimmermann, L. and Wagner , M. (2025). 'Mapping the chemical complexity of plastics'. *Nature* 643: 349–355. https://doi.org/10.1038/s41586-025-09184-8.

Mori, A.S., Furukawa, T. and Sasaki, T. (2013). 'Response diversity determines the resilience of ecosystems to environmental change'. *Biological Reviews* 88: 349–364. https://doi.org/10.1111/brv.12004.

Mori, A.S., Isbell, F. and Cadotte, M.W. (2023). 'Assessing the importance of species and their assemblages for the biodiversity-ecosystem multifunctionality relationship'. *Ecology* 104: e4104. https://doi.org/10.1002/ecy.4104.

NatureScot. (2020). 'Natural capital, ecosystem services and the blue economy'. Retrieved from https://marine.gov.scot/sma/sites/default/files/hns_09_natural_capital.pdf (accessed 22 May 2026).

Nicotra, A.B., et al. (2010). 'Plant phenotypic plasticity in a changing climate'. *Trends in Plant Science* 15: 684–692. https://doi.org/10.1016/j.tplants.2010.09.008.

Okello, C., Githiora, Y.W., Sithole, S. and Owuor, M.A. (2024). 'Nature-based solutions for water resource management in Africa's arid and sem-arid lands (ASALs): a systematic review of existing interventions'. *Nature-Based Solutions* 6: 100172. https://doi.org/10.1016/j.nbsj.2024.100172.

Peterson, G.D. (2008). 'Self-organization and discontinuities in ecosystems'. In C.R. Allen and C.S. Holling (eds), *Discontinuities in Ecosystems and Other Complex Systems*. New York: Columbia University Press.

Pimm, S.L. (1982). *Food webs.* London: Chapman and Hall.

Ramirez-Llodra, E. and Keith, D.A. (2020). 'M3.7 Chemosynthetic-based-ecosystems (CBE)'. In D.A. Keith, J.R. Ferrer-Paris, E. Nicholson and R.T. Kingsford (eds), *IUCN Global Ecosystem Typology 2.0: Descriptive Profiles for Biomes and Ecosystem Functional Groups*. Gland, Switzerland: IUCN. https://doi.org/10.2305/IUCN.CH.2020.13.en.

Resing, J.A., Sedwick, P.N., German, C.R., Jenkins, W.J., Moffett, J.W., Sohst, B.M. and Tagliabue, A. (2015). 'Basin-scale transport of hydrothermal dissolved metals across the South Pacific Ocean'. *Nature* 523: 200–203. https://doi.org/10.1038/nature14577.

Ryan, M.G., Binkley, D. and Fownes, H. (1997). 'Age-related decline in forest productivity: pattern and process'. *Advances in Ecological Research* 27: 213–262. https://doi.org/10.1016/S0065-2504(08)60009-4.

Silvy, N.J. (ed.). (2012). *The Wildlife Techniques Manual: Research*, vol. 1 (7th edition). Baltimore, ND: Johns Hopkins University Press.

Standish, R.J. and Parkhurst, T. (2024). 'Interventions for resilient nature-based solutions: an ecological perspective'. *Journal of Ecology* 112: 2502–2509. https://doi.org/10.1111/1365-2745.14377.

Stokes, B.R. and Keitt, T.H. (2025). 'An intergeneric hybrid between historically isolated temperate and tropical jays following recent range expansion'. *Nature Notes* 15(9): e72148. https://doi.org/10.1002/ece3.72148.

Sutter, P. (2023). 'How did the universe's elements form?' Retrieved from www.space.com/how-did-the-universe-elements-form (accessed 8 September 2024).

Tansley, A.G. (1935). 'The use and abuse of vegetational concepts and terms'. *Ecology* 16: 284–307.

ten Hietbrink, S., Materić, D., Holzinger, R., Groeskamp. S. and Niemann, H. (2025). 'Nanoplastic concentrations across the North Atlantic'. *Nature* 643: 412–416. https://doi.org/10.1038/s41586-025-09218-1.

Thapa, R., Thoms, M. and Parsons, M. (2016). 'An adaptive cycle hypothesis of semi-arid floodplain vegetation productivity in dry and wet resource states'. *Ecohydrology* 9: 39–51. https://doi.org/10.1002/eco.1609.

Tran, H.T., et al. (2022). 'Phthalates in the environment: characteristics, fate and transport, and advanced wastewater treatment technologies'. *Bioresource Technology* 126249. https://doi.org/10.1016/j.biortech.2021.126249.

Tzoulas, K., Galan, J., Venn, S., Dennis, M., Pedroli, B., Mishra, H., Hasse, D., Pauleit, S., Niemelä, J. and James, P. (2021). 'A conceptual model of the social-ecological system of nature-based solutions in urban environment'. *Ambio* 50(2): 335–345. https://doi.org/10.1007/s13280-020-01380-2.

United Nations. (2011). *Convention on Biological Diversity: Text and Annexes.* Montreal, Canada: Secretariat of the Convention on Biological Diversity, United Nations Environment Programme.

Varjani, S., Joshi, R., Srivastava, V.K., Ngo, H.H. and Guo, W. (2020). 'Treatment of wastewater from petroleum industry: current practices and perspectives'. *Environmental Science and Pollution Research* 27: 27172–27180. https://doi.org/10.1016/j.biortech.2021.126059.

Walker, B., et al. (2023). 'Response diversity as a sustainability strategy'. *Nature Sustainability* 6: 621–629. https://doi.org/10.1038/s41893-022-01048-7.

Whittaker, R.H. and Likens, G.E. (1973). 'Primary production: the biosphere and man'. *Human Ecology* 1: 357–369. https://doi.org/10.1007/BF01536732.

Winter, K.B., Lincoln, N.K. and Berkes, F. (2018). 'The social-ecological keystone concept: a quantifiable metaphor for understanding the structure, function, and resilience of a biocultural system'. *Sustainability* 10(9): 3294. https://doi.org/10.3390/su10093294.

Yachi, S. and Loreau, M. (1999). 'Biodiversity and ecosystem productivity in a fluctuating environment: the insurance hypothesis'. *PNAS* 96: 1463–1468. https://doi.org/10.1073/pnas.96.4.1463.

Zavaleta, E., Pasari, J.R., Hulvey, K.B. and Tilman, G.D. (2010). 'Sustaining multiple ecosystem functions in grassland communities requires higher biodiversity'. *PNAS* 107: 1443–1446. https://doi.org/10.1073/pnas.0906829107.

7

LANDSCAPE ECOLOGY

7.1 INTRODUCTION

Ecologically, a landscape comprises all the ecosystems (see Chapter 6) occupying the same space. Landscape ecology provides an understanding of how ecosystems function alongside each other across different temporal and spatial scales and how human activities impact the functions. Landscape ecologists consider the spatial and temporal dynamics of ecosystems (e.g. the spatial arrangement of ecosystems across a landscape and the movement of organisms (individuals and populations), materials, and energy between these ecosystems, and how these spatial arrangements and flows change over time). They also seek to understand the impacts of human activities on landscapes and how modifications in landscape patterns affect ecosystem services such as purifying water, regulating climate, and support biodiversity.

The first main section in this chapter begins with a definition of a landscape from the perspective of a landscape ecologist. The focus then shifts to the three important theoretical foundations of landscape ecology: island biogeography theory, metapopulation theory, and patch–corridor–matrix models. Each of these is discussed. In the same section there is also a consideration of boundaries (with a focus on edges) and an outline of the guiding principles and core concepts of landscape ecology.

In the third section in this chapter there is a discussion of the fragmentation-biodiversity debate. This debate is relevant, for example, when considering how nature reserve should be placed across a landscape to achieve the best result for biodiversity conservation.

DOI: 10.4324/9781003604464-7

The issue can be phrased simply as, is it better to have a single large area of habitat or several smaller areas of the same habitat which combined have the same overall area as the single, large patch.

Using their insights, landscape ecologists contribute to multi- and inter-disciplinary programmes to sustainably manage landscapes in order to maintain or restore landscape-scale ecological processes, such connecting rivers to their flood plains or providing corridors through which animals can migrate. In so doing, landscape ecologists contribute to actions including those addressing the effects of climate change and biodiversity loss.

The case study on tackling habitat loss brings together the fragmentation-biodiversity debate, with a knowledge and understanding of how landscapes are structured and function to address the global issue of habitat loss.

7.2 LANDSCAPES FROM AN ECOLOGICAL PERSPECTIVE

In ecological terms, a landscape is a spatially heterogeneous area composed of interacting ecosystems (or patches). The structure and function of a landscape is shaped by both natural processes and human activities. Landscape ecology, a term originally coined by Carol Troll (1899–1975, a German geographer) in 1939, explicitly recognises the diversity and arrangement of land units (e.g. forests, grasslands, wetlands, agricultural fields, and urban areas) and the dynamic interactions among them. This view is often contrasted with that based on the traditional ecosystem concept which often assumes spatial homogeneity and focuses on discrete units. A spatial perspective is central to landscape ecology. A landscape is not defined by a fixed size; rather, its scale is determined by the ecological phenomenon or organism under consideration. This means that landscape ecological principles and ideas can be applied at different spatial scales including the home range of a small invertebrate which may be a few square metres, to the catchment area of a river system or to transcontinental migratory corridors. Landscape ecologists investigate how spatial patterns (e.g. the distribution, size, and connectivity of patches) influence ecological processes, biodiversity, and the provision of ecosystem services (i.e. the benefits that

Figure 7.1 The landscape of the Zanskar Valley in the high Himalayas of Ladakh, India during the summer illustrating landscape structure (e.g. mountains, valleys, settlements, vegetated and unvegetated areas), function/processes (e.g. flow of eroded material from the mountain into the valley and erosive effects of the flow of water in the river) and change (changes in the size of the settlement and changes in the snow level at different times of the year.

Source: photograph by Rob Young

humans derive from ecosystems, essential for well-being, economic activity, and environmental sustainability).

Ecologically, landscapes (Figure 7.1) are viewed as having three main attributes: structure, function/processes, and change. These attributes are used as the basis for quantifying and comparing landscapes (Table 7.1). As landscapes change so do these three attributes and, in turn landscape process (including flows of materials, nutrients, and organisms) are disrupted to a lesser or greater extent. Disruption of these flows affects the maintenance of critical ecological functions that contribute to the structure and diversity (i.e. heterogeneity) of a landscape.

There are three dominant theories used in the ecological study of landscapes. In 1967, Robert MacArthur (1930–1972, a

Table 7.1 The three main attributes of a landscape

Landscape attribute	*Description*	*Examples in Figure 7.2*
Structure	The diversity (shape, size, and composition) and spatial arrangement of landscape elements (e.g. habitat patches).	Shape and size of the mountains Size of the rivers Size and shape of the blocks of vegetation Size and shape (including variations within) of villages, towns, and cities (see Chapter 3)
Function/ Processes	The interaction among the spatial elements that make up a landscape (e.g. the flow of energy, nutrients, species, or genes among habitat patches; small population dynamics/genetics; species interactions/ community dynamics; patch heterogeneity; species dispersal).	Flow of water from the mountains into the river Flow of rocks from the mountains into the plain Flow of air resulting from the cooling effect of the snow on the mountains and warmer air rising from the valley Flow of energy associated with the movement of water in the river
Change	The variation in landscape structure and function/ processes over time.	Spread or retraction of the villages according to population size Spread or reduction of the cultivated areas Changes in the position of the river as the water erodes the banks

Canadia-born American who made major contributions to population and community ecology) and Edward O. Wilson (1929–2021, an American biologist, naturalist, ecologist, and entomologist) published the theory of island biogeography (MacArthur and Wilson 1967). Two years later Richard Levins (1930–2016), an American biologist, population geneticist, biomathematician, mathematical ecologist, and philosopher of science, coined the term 'metapopulation' (Levins 1969), and patch–corridor–matrix models were developed by Richard Forman (1935–), an American landscape ecologist, and Michel Godron (1931–), a French ecologist (Forman 1981; Forman and Godron 1981).

7.2.1 ISLAND BIOGEOGRAPHY THEORY

Early naturalist-explorers and biogeographers noted that larger islands held more species than smaller islands. In 1957, P.J. Darlington Jr. (1904–1983, an American entomologist, field naturalist, biogeographer, and zoologist) suggested a rule of thumb (sometimes referred to as Darlington's rule): a tenfold increase in area leads to a doubling in the number of species (Darlington 1957). This rule has stood the test of time fairly well, but isolation of the island from the mainland can lead to significant exceptions. Five years later F.W. Preston (1896–1989, an English American engineer, ecologist, and conservationist) formalised the relationship between the area of an island and the number of species present (Preston 1962) as:

$$S = cA^2$$

or

$$\log S = \log c + z \log A$$

where S = the number of species, A = the area of the island, c = a constant related to the number of species per unit area, log = logarithm to base 10, and z = a constant which is a measure of the change in species richness per unit area obtained by measuring the slope of the line relating S and A.

The second equation explaining this relationship is known as a 'log-log' equation, in which both sides of the equation are transformed on to a logarithmic scale. This transformation is often used when the data spans several orders of magnitude (e.g. from tens to millions of square metres). Plotting a 'log transformed' curve effectively converts the nonlinear relationship to a linear curve (i.e. a straight line) and aids further analysis (Figure 7.2). The actual shape of the curve and slope of the straight line are dependent on the two variables c and z.

The central tenet of island biogeography is that the biota on an island maintain a dynamic equilibrium between immigration and extinction rates and that these rates depend on the size and isolation of the island from source regions (MacArthur and Wilson 1967). In the simple example in Figure 7.3 there are three islands of approximately equal size, located at different distances from the

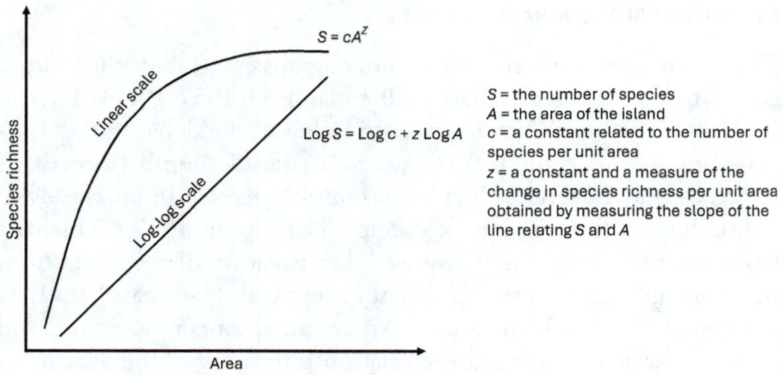

Figure 7.2 Graphical representation of the relationship between species richness and area.

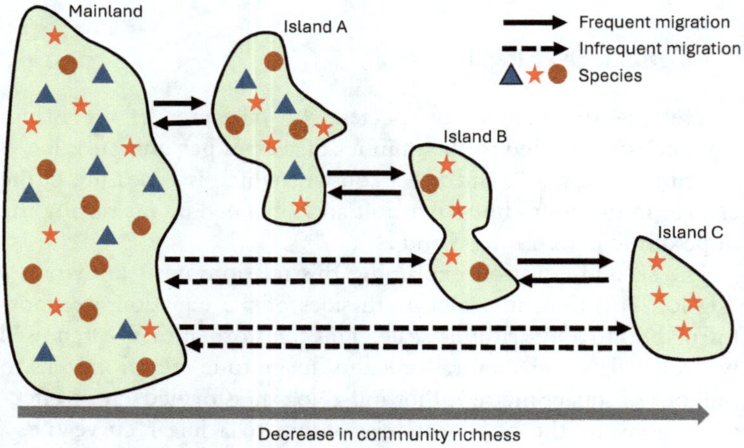

Figure 7.3 Community richness is related to the isolation of the island from the mainland.

mainland. The extinction rate of species is assumed to be the same on each island. Community richness, therefore, is related only to immigration from the mainland or other islands. The result is that islands that are further away from the source community (the mainland) have a lower community richness than those which are nearer.

In this simplified model there is an assumption that extinction rates are constant, but this is not always valid as rates commonly vary between islands.

The dynamic equilibrium between immigration and extinction is frequently presented as a graph with two lines. One line for immigration rate of new species and the other for extinction rate of resident species versus the number of species present on an island. The intersection point of these two lines gives the equilibrium species number (Figure 7.4). The rate at which one species is lost and a replacement gained is called the turnover rate.

Immigration, extinction, and turnover rates and, consequently, the predicted equilibrium species richness are influenced by the size and proximity of an island to the mainland. It can be hypothesised that islands near to the mainland will reach equilibrium with more species than those that are distant, and small islands will reach equilibrium with fewer species than large islands. Turnover rates are greater for near islands than for distant ones of the same size, but turnover rates for near, small islands are greater than for near, large islands because extinction rates are greatest on small islands (Figure 7.5).

The equilibrium theory of island biogeography focuses on species richness only, treating all species as equivalents and making no

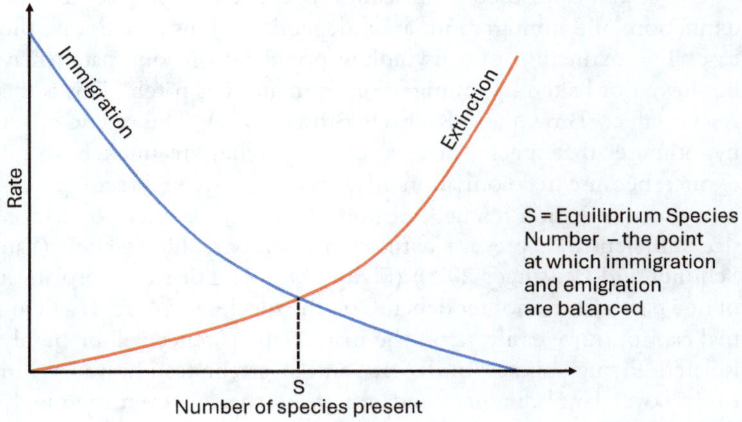

Figure 7.4 The MacArthur and Wilson (1963) equilibrium model.

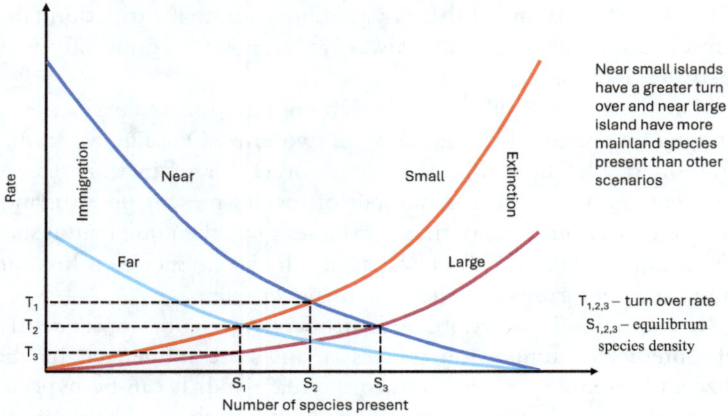

Figure 7.5 The influence of size (small/large) and proximity (near/far) of an island to the mainland on species turn-over rate and species equilibrium density.

assumptions about species composition. The theory also assumes that the probabilities of immigration and extinction are the same for all species. This assumption does not take account of limitations related to the life history or habitat requirements of the species (e.g. birds migrate more easily than mammals).

The equilibrium theory of island biogeography also assumes that extinctions and immigrations are independent. This may not be the case. The extinction of a dwindling population in one patch may be slowed or halted by immigration from another patch. This is the rescue effect (Brown and Kodrich-Brown 1977). The rescue effect hypothesises that species in less isolated patches are unlikely to go extinct because recolonisation may occur between breeding seasons ('recolonisation rescue'), or immigrants may bolster population size sufficiently to prevent extinction ('demographic rescue') (Van Schmidt and Beissinger 2020) (Figure 7.6). The degree of isolation of one patch from another depends on the species involved. If an animal cannot move easily across the matrix, the patches will be highly isolated. Flying animals (birds, bats, and invertebrates), however, can easily cover large distances and, for those species, even seemingly distant patches may not be isolated to any meaningful degree.

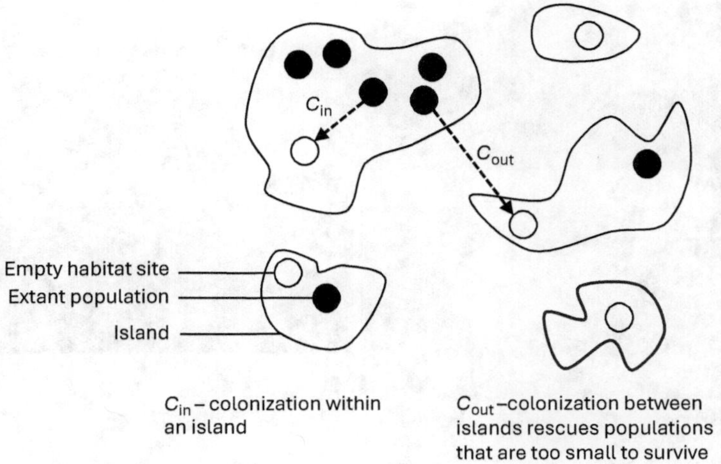

Figure 7.6 The rescue effect on islands or between patches of habitat.

Since its initial focus on islands, the theory of island biogeography has been extended to terrestrial environments as well as to issues such as host plants and their insects, and to animals and their parasites. These extensions do not come without issues. Islands in an ocean and terrestrial patches differ in significant ways (Figure 7.7). Water is, generally, an inhospitable habitat for terrestrial animals. To cross from one island to another species must fly, be transported by flying animals, carried by the wind or ocean currents, or swim (which not all terrestrial animals can do). In the terrestrial environment, though the habitat surrounding an island may be very different from that of the island (e.g. grass surrounding islands of trees), it is generally less inhospitable than water for terrestrial organisms. To cross from, say, one woodland to another, animals may, according to the species, use the same aerial routes as those crossing from one oceanic island, but instead of swimming, they can run, walk, hop, or crawl. Furthermore, oceanic islands tend to be larger and have greater diversity and more species richness than relatively smaller, homogeneous, terrestrial habitat patches which are, consequentially, occupied by relatively few species.

Figure 7.7 Comparison of 'islands' in aquatic and terrestrial environments. (A) Islands surrounded by sea near Stockholm, Sweden. (B) 'Islands' of trees in grassland in Cheshire, England. (C) 'Islands' of lakes and in a forest in Nahuel Huapi National Park, North Patagonia, Chile.

Sources: (A) photograph by Matthew Dennis; (B) photograph by P. James; (C) photograph by Jan Sayle

Habitat diversity theory, which states that it is the diversity of habitats that supports species richness (not area *per se*), is an alternative to island biogeography. There is evidence that habitat heterogeneity can override the influence of island size, with smaller islands having high habitat heterogeneity supporting a greater species richness than larger, more homogeneous islands (Rigby and Lawton 1981; Blake and Karr 1984; Freemark and Merriam 1986).

7.2.2 METAPOPULATION THEORY

As noted earlier, the word 'metapopulation' was coined by Richard Levins in 1969. A metapopulation consists of a group of spatially separated populations of the same species. Since its initial conception, the theory has been applied to species in naturally or artificially fragmented habitats in both aquatic and terrestrial ecosystems (Kritzer and Sale 2006).

In classical metapopulation theory, each population in its habitat patch is relatively independent of other populations in different habitat patches in terms of its establishment, growth, decline, and eventual extinction as a consequence of demographic stochasticity (i.e. fluctuations in population size due to random demographic events) (Figure 7.8). Within this model, some habitat patches that could be occupied by a population may, at any one time, be unoccupied.

Although individual populations in individual habitat patches have finite lifespans, the metapopulation is often stable. Immigration from an extant population is likely to lead to the recolonisation of an empty habitat patch or bolster a population which had become low and rescue it from extinction. In the case where a population is below the carrying capacity (see Chapter 4), there will be niche opportunities open to 'rescuers' from another population. This is another example of the rescue effect (see Figure 7.6 and accompanying text).

The development of metapopulation theory, in conjunction with that of source-sink, emphasised the importance of connectivity between seemingly isolated populations. Source-sink dynamics is a theoretical model to describe how variation in habitat quality may

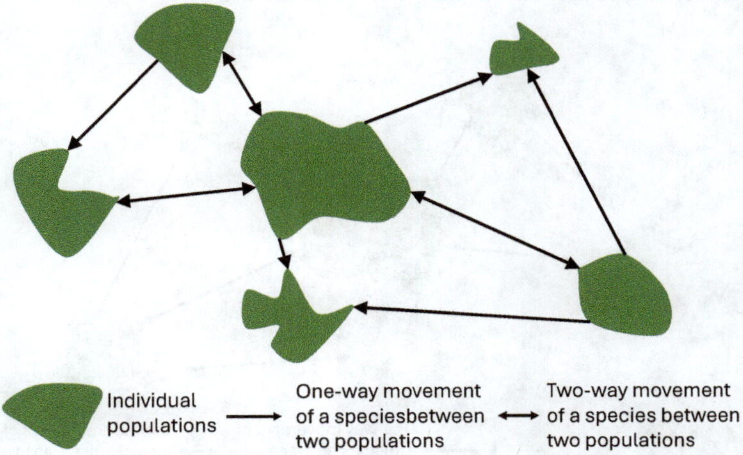

Figure 7.8 The classical model of a metapopulation.

affect population growth or decline. In the simplest form of this model, organisms occupy two patches of habitat. One patch, the source, is a high-quality habitat that, on average, allows the population to increase. The second patch, the sink, is a very low-quality habitat that, on its own, would not be able to support a population. If excess individuals in the source population (i.e. those beyond carrying capacity of the source) move frequently from the source to the sink, the sink population can persist indefinitely. This simple model can be extended to cases where there are more than one source populations and more than one sink habitat (Figure 7.9).

Movement between source and sink is affected by the distance between the sources and sink. As the distance increases, so the frequency of individuals moving between the source and sink and the total number of individuals that make the trip decrease (Figure 7.10).

Although the long-term survival of a given single population may not be guaranteed, the combined effect of movements between populations within a metapopulation may ensure the long-term survival of the species. The source–sink model implies that some habitat patches may be more important to the long-term survival of the population than other patches, and may be important in informing conservation management plans and actions.

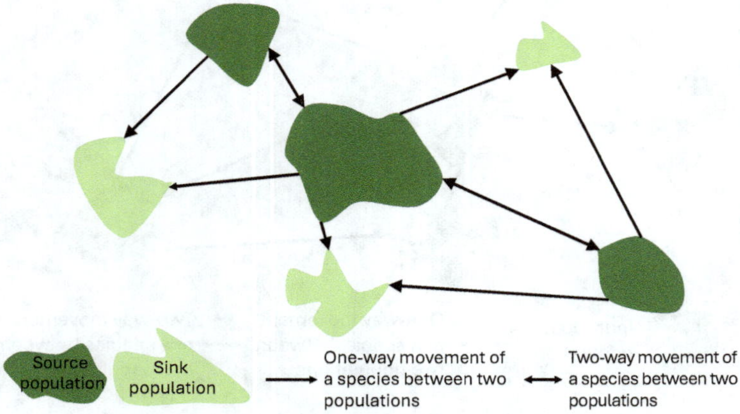

Figure 7.9 The source–sink model of metapopulations.

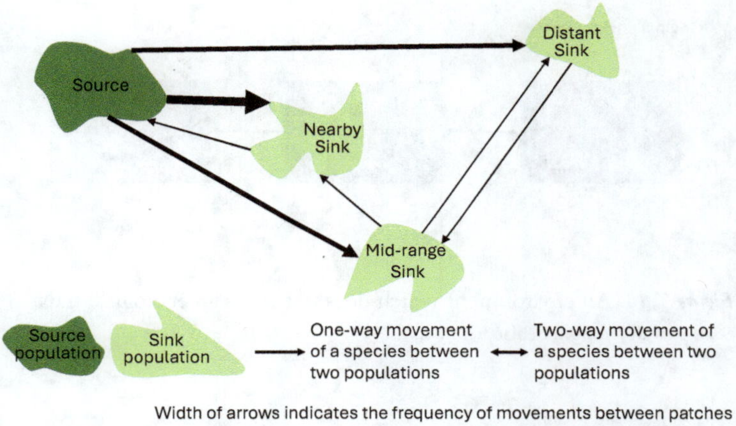

Figure 7.10 The rescue effect operating at different rates according to the distance from the source population to different sink populations.

In the source-sink model it is generally assumed that organisms can distinguish between high- and low-quality habitat and prefer high-quality habitat. Such a choice may not always be soundly based. Where inappropriate decisions are made, this is termed an 'ecological trap'. Ecological traps occur when either the habitat quality or the cue(s) change so that organisms occupy a poor-quality habitat. A cue, in this sense, is an environmental factor that informs and influences an animal's behaviour. Normally, a cue in the original environment of an organism elicits a corresponding behavioural response associated with an appropriate predictable, adaptive outcome (Schlaepfer et al. 2002). For example, in spring a bare patch of soil within a grassland is a cue to the ground nesting northern lapwing (*Vanellus vanellus*) that it is an ideal place in which to lay eggs and successfully raise its young (Figure 7.11).

An ecological trap can arise in either of two ways. The first is the result of environmental change. In this instance a cue occurs in an environment which is no longer appropriate for the resultant behaviour (Schlaepfer et al. 2002). To continue with the example of the ground nesting bird, the grassland is still there, the bare patch is still there, but there is now a new predator in the area

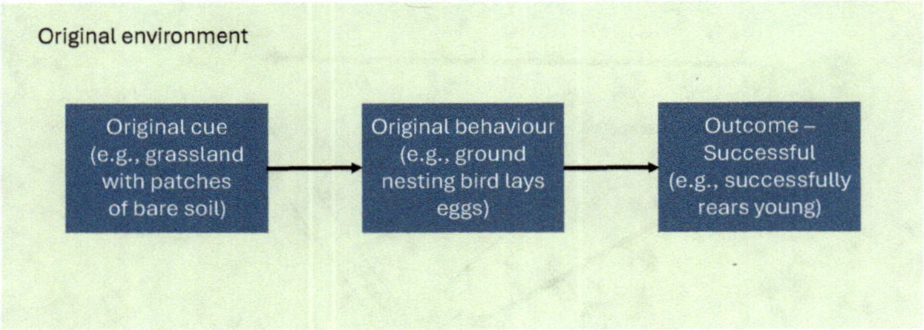

Figure 7.11 An environment which does not act as an ecological trap.
Source: modified from Schlaepfer et al. (2002)

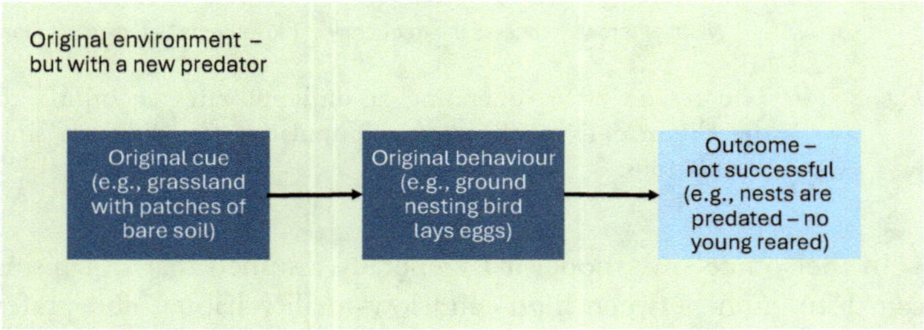

Figure 7.12 Ecological trap arises due to a new factor in the environment that is not perceived by the organism.
Source: modified from Schlaepfer et al. (2002)

which take the eggs once they are laid (i.e. the environment has changed) (Figure 7.12). For example, the stoat (*Mustela erminea*) was introduced to Aotearoa/New Zealand in 1879 to control the European rabbit (*Oryctolagus cuniculus*) which had been introduced in the 1830s (i.e. the environment changed). Stoats are voracious and relentless hunters and are known to have had a significant effect on bird species by taking eggs, chicks, and adults. In response, the Aotearoa/New Zealand Predator Free 2050 initiative aims to eradicate introduced predators including stoat, rats (*Rattus* spp.), and possums (Phalangeriformes) from the country by 2050.

The second way in which an ecological trap occurs is when the altered environment contains a novel element that does not provide the resources needed but mimics the original cue closely enough to

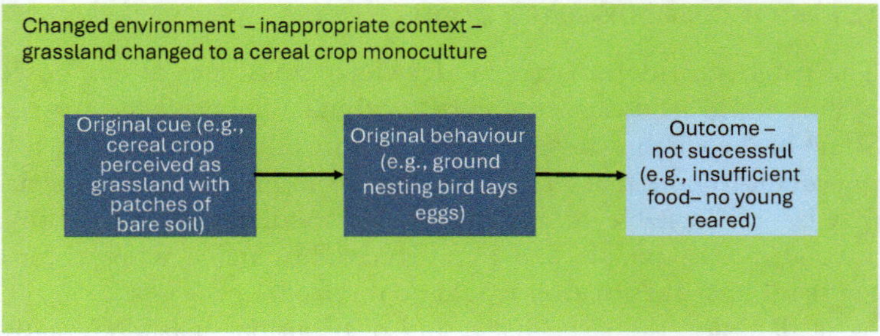

Figure 7.13 Ecological trap that arises from a change in the environment that mimics the original environment.

Source: modified from Schlaepfer et al. (2002)

elicit the original behaviour (Schlaepfer et al. 2002) (Figure 7.13). To complete the ground nesting bird example, this type of ecological trap occurs when, say, a grassland is turned into an agricultural field sown with a cereal crop. As cereals are grasses, the ground nesting bird will perceive a field of grass with some bare patches and accordingly lay eggs. However, that cereal field may not contain a sufficient insect population, compared with the original grassland, to support the successful rearing of any young birds. Another example of this type of trap is seen in the behaviour of female mayflies (Ephemeroptera) and dragonflies (Odonata). The normal behaviour of the females is to lay their eggs in waterbodies. The environmental cue used to locate suitable waterbodies is reflected polarised light. Asphalt roads reflect polarised light in a similar way to a waterbody. As a result, female mayflies, dragonflies, and similar aquatic invertebrates lay their eggs on the asphalt. Brown pelican (*Pelecanus occidentalis*), like aquatic invertebrates, perceive polarised light and mistake an asphalt surface for a water body. They land, mistakenly, on the asphalt and in so doing frequently injure themselves. Marine animals are also at risk from in appropriate polarised light cues. Plastic reflects polarised light. It is thought that this results in marine turtles mistaking plastic for jellyfish, their preferred food source. The effect of eating plastic on the turtles includes damage to their internal organs and reduced ability to and digest other food sources, thereby reducing their long-term growth and survival (Brayley 2022).

7.2.3 PATCH–CORRIDOR–MATRIX MODEL

The patch–corridor–matrix model views landscapes as composed of 'hospitable' landscape patches set within an 'inhospitable' matrix. Some of the patches are connected by corridors (Figure 7.14). It is important to note that a landscape that is 'hospitable' to one species may be 'inhospitable' to another (e.g. a woodland bird would find a woodland 'hospitable' and a grassland 'inhospitable' whereas for a grassland bird the situation would be reversed).

At all scales, the size, shape, and distribution of patches across a landscape affect colonisation by individuals and the persistence of individuals and populations in the landscape. Examples of the importance of patchiness at different scales can be seen in two examples. At a small scale, the size and number of patches of goldenrod plants (*Solidago* spp.) in a field is important to the goldenrod beetle (*Diabrotica* spp.). At a much larger scale, large patches of woodland and grassland are important for deer (Cervidae).

Natural and human disturbances also change the size of patches There is a critical size below which a patch cannot support all the species that were originally there (Whitcomb et al. 1976; Karr 1982; Robbins et al. 1989). The actual size of that critical point depends on the type of patch (e.g. it will be different for a patch of woodland

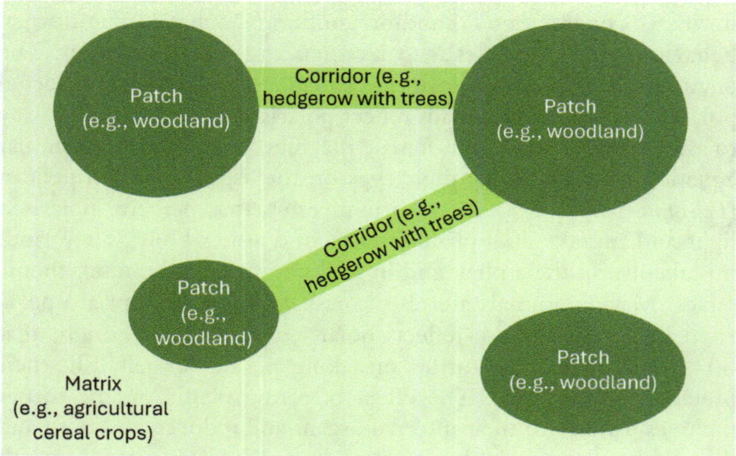

Figure 7.14 The patch–corridor–matrix model of a landscape.

than for an area of grassland). The first set of species to disappear from the community as the size of a patch decreases comprise those that require large territories or foraging areas and interior species (so named as their habitat does not include habit edges).

The minimum size of habitat needed to maintain interior species differs between individual species. For interior plant species survival, patch area, *per se*, is not as important as environmental factors such as moisture and light. That said, there appears to be a general pattern of maximum species diversity occurring in patches of intermediate size. This results from a combination of a positive correlation between interior species and increased area and an inverse corelation between edge species and the size of the habitat patches (Weaver and Kellman 1981).

The concept of a patch becomes more complicated when the heterogeneity and dynamics of the landscape are considered. For example, not all patches of woodland are equal in terms of their hospitality to woodland core birds and all woodlands in a landscape will not be the same age or at the same stage in succession. Similarly, not all ponds are equally hospitable to aquatic plants and animals. Ponds vary in multiple ways including depth, topography, substrate, water chemistry, amount of shade, and age. This means that there is a heterogenous group of patches across a landscape.

The same complications arise over the concept of a corridor. Corridors link patches of similar habitat, for example a corridor of trees and shrub linking two woodlands. Some corridors are narrow; others are wide. In the example of a corridor linking two woodlands, narrow corridors may be lines of trees planted as windbreaks, hedgerows, roadside verges, and drainage ditches (Forman 1986, 1995). Wide corridors, often called strip corridors (e.g. a wide strip of woodland), have both interior and edge environments. They may also serve as human rights-of-way or include powerlines or other services.

Corridors can act as dispersal routes and travel lanes for species moving between habitat patches (Rosenberg et al. 1997). This can facilitate gene flow between populations (and sub-populations) in those different patches. Such movements may facilitate the re-establishment of populations in patches in which they have become locally extinct (i.e. the rescue effect). However, predators and disease organisms may also travel along corridors and, hence, gain access to prey/hosts in different patches.

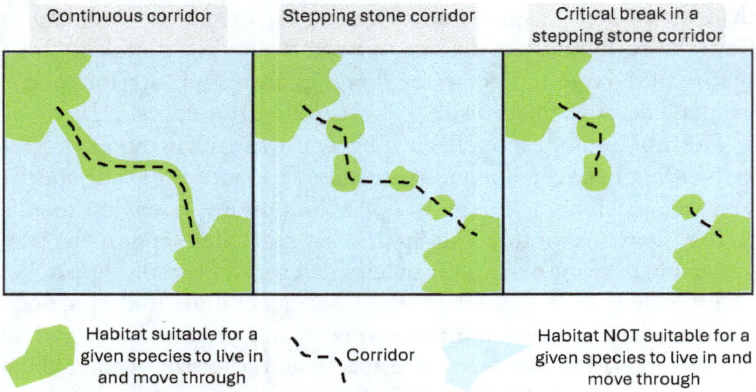

Figure 7.15 Continuous and stepping stone corridors connecting two large habitat patches.

What is a corridor for one species may not have that function for another. Corridors can act as filters, providing dispersal routes for some species but not for others. Gaps of various sizes in the corridor can filter those species that are able to cross the gap from those which are less able. For example, in a forest environment the Japanese giant flying squirrel (*Petaurista leucogenys*) has been observed to cross a 30 m gap (or there abouts) whereas a stick insect (*Aretaon asperrimus*) can only cross a gap of around 50 mm (Graham and Socha 2018).

Corridors for highly mobile species need not necessarily be continuous; they may be stepping-stones set within the matrix providing places where highly mobile species can rest and feed before moving on to the next patch. A critical gap can occur when the distance between adjacent stepping stones is too large for the organism to cross (Figure 7.15). Establishing functional corridors, including installing a stepping stone in a critical gap, is an approach used in conservation management.

7.2.4 BOUNDARIES

One of the most conspicuous features in a landscape are the boundaries between one landscape element and another, say a woodland and grassland, or grassland and a pond. Some boundaries are inherent and stable. These are the result of abrupt changes in soil type, topography, geomorphic features (such as rock outcrops), and micro-climate. Others, are induced, periodic, and ephemeral, for example, those created by fires, storms, and floods or resulting from human-induced disturbance such as livestock grazing, timber harvesting, agriculture, and urbanisation. Without periodic disturbance, an induced edge will revert to its pre-disturbance state. For example, if trees are felled around the outsides of a woodland, a process of succession will take place which, over time, results in the re-establishment of the former woodland. During this re-establishment, the character of the edge will change. Firstly, trees on a newly created edge, due to more exposure to light, are likely to sprout branches (termed epicormic branches) on the lower parts of their trunks. Later, shrubs and small trees will begin to grow where trees had been felled, and succession will continue until the woodland is re-established.

There is frequently a transition zone (called an ecotone) between the two habitats at either side of the edge. The species found in an ecotone may be a mixture of those found in the two adjoining habitats. For example, where a forest merges into grassland, the species present will comprise those woodland edge species and grassland edge species which are adapted to survive where there is some shade (Figure 7.16).

There are three types of edge effects which influence the characteristics of the edge (Murcia 1995) (Table 7.2). The extent of the edge effect is influenced by the length and width of the edge and by the degree of contrast between adjacent habitats (Patton 1975; Sisk et al. 1997).

Edges also differentially facilitate or restrict the dispersal of plants and animal movement, thus altering species interactions (Forman 1995; Sisk et al. 1997; Farina 1998). For example, the growth of thorny plants and dense low growth can inhibit the movement of animals. A second example is the presence of domesticated animal predators (e.g. cats and dogs), herbivores (e.g. sheep, goats, and cattle), and species that are highly adapted to living with humans. These species can and do move into the surrounding remnant core habitat and in so doing increase the predatory pressure on plants and animals in those interiors.

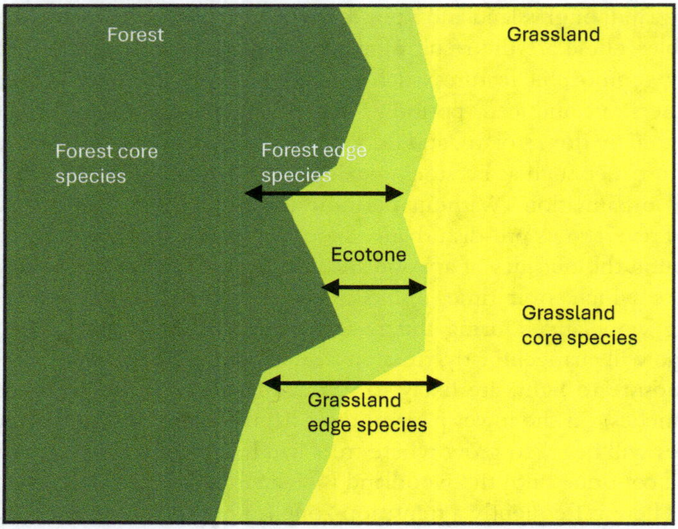

Figure 7.16 A forest–grassland ecotone (or transition zone).

Table 7.2 Edge effects on habitat fragments

Type of edge effect	*Description*
Abiotic effects	Changes in environmental conditions resulting from differences on either side of the edge.
Direct biological effects	Changes in the abundance and distribution of species, either as a direct result of altered abiotic conditions or indirectly, due to the physiological tolerances of species to conditions at and near the edge (e.g. higher light levels, wind exposure, temperatures, and humidity).
Indirect biological effects	Changes in species interactions, such as predation, brood parasitism, competition and herbivory, biotic pollination, and seed dispersal.

Source: Murcia (1995)

7.2.5 GUIDING PRINCIPLES AND CORE CONCEPTS

Emerging from island biogeography theory, metapopulation theory, and patch–corridor–matrix models are several principles or core concepts that guide landscape ecology research and practice (Table 7.3).

Table 7.3 Guiding principles and core concepts in landscape ecology

Guiding principles or core concepts and comments
Landscapes are heterogeneous
Landscapes are mosaics of different types and quality of landforms, ecosystems, habitat types, or land uses. While landscapes are often viewed as comprising discrete elements (e.g. habitat patches, corridors, roads, water bodies), heterogeneity can vary continuously over the landscape (e.g. along an ecocline or ecological gradient).
Landscapes are diverse in form and function
Landscapes contain a diverse array of features related to their geomorphology, primary land cover, or land use. They have ecological or biological functions and provide amenities or commodities (e.g. food production) and values (e.g. cultural landscapes).
Landscapes are scale-dependent
While landscapes have traditionally been viewed on human terms, as areas of broad spatial extent, they can also be viewed as a spatially heterogeneous area. This allows landscapes to be viewed at different scales according to the species, according to the distribution and availability of their resource requirements, and the different scales at which they operate.
Landscapes are dynamic
All ecological systems, including landscapes, are dynamic. Disturbances operate across a wide range of temporal and spatial scales, from those occurring over minutes within a few square centimetres, to those operating over tens of millennia across thousands of square kilometres. Landscapes are best viewed as shifting mosaics.
Spatial context is important
Ecological dynamics vary spatially. For example, population growth rates of a given species vary according to the suitability of habitats for that species.

(*Continued*)

Table 7.3 (Continued)

Guiding principles or core concepts and comments
Ecological flows are important
The mosaic of land cover or land use types that occur between habitat patches can influence the magnitude of ecological flows among patches. Patch boundaries may be hard (if organisms are unwilling or unable to cross that boundary) or soft (where there is a more gradual transition from one habitat type to the next). Two-way flows of materials (including individuals, genes, stormwater runoff, nutrients, and pollutants) also occur between habitat patches.
Connectivity is important
Ecosystems can be more adaptive and resilient when connected. Connectivity is an emergent property of landscapes – the consequence of the interaction between ecological flows and the landscape pattern. Connectivity is important for understanding the propagation of disturbances across the landscape; the movement and redistribution of organisms, materials, and nutrients; the resulting structure and dynamics of populations; gene flow and population genetic structure; the spread of invasive species and diseases; community patterns and dynamics; and ecosystem structure and function.
Landscapes are multifunctional
Landscapes provide humanity with an array of goods and services which can lead to competing interests and valuation systems. Sustainability is a key concept in landscape ecology and ecosystem and natural resource management.

Source: based on With (2019)

7.3 FRAGMENTATION-BIODIVERSITY DEBATE

Fragmentation is a process in which an area of habitat becomes divided into smaller, isolated areas but where the overall area of habitat remains the same. The process of fragmentation can be summarised in four phases: perforation (initial small openings) (Figure 7.17A), dissection (larger intrusions of change, often along physical features) (Figure 7.17B), dissipation (spread and coalescing of alteration) (Figure 7.17C), and eventually shrinkage (reduction of patch size) and attrition (loss of patches) (Figure 7.17D).

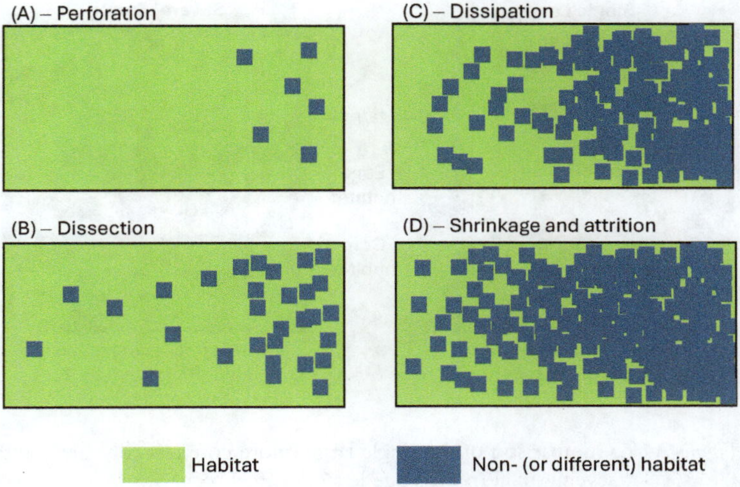

Figure 7.17 Process of landscape fragmentation.
Source: adapted from Clark (2010)

The size of most populations is adversely affected by habitat fragmentation. The exceptions are sedentary species (e.g. some woodland plants) which are adversely affected only if their particular location is destroyed, and species with very good dispersal mechanisms (e.g. plants with air born seeds, large birds, small flying mammals, and larger mammals) which can be adversely affected by a decrease in the total amount of habitat available.

To address issues of fragmentation some conservation interventions are focused on conserving remaining habitat patches and restoring functional landscapes. A question this raises is which is the more appropriate, to conserve or restore a single large (SL) patch or several small (SS) patches? This is often referred to as the SLOSS (single large or several small) debate or, in more up to date terminology, the fragmentation-biodiversity debate.

One relevant principle, inspired by the theory of island biogeography, is that a single large reserve (SL) should hold more species than several small reserves (SS) of the same total area: this is the 'SL > SS principle' (MacArthur and Wilson 1963, 1967; see also May 1975; Diamond 1976). If the sum of the overall area of

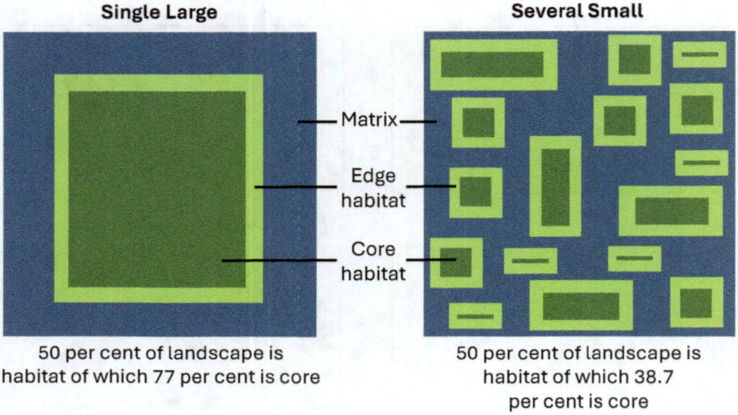

Figure 7.18 Comparison of the single large (more core less edge area) and several small (more edge less core area) scenarios.

a single large or several small patches is the same, there will be more edges and area of ecotone habitat (an ecotone is the transitional area between two different ecosystems or biomes) in the several small scenario, and this will lead to greater overall biodiversity (Figure 7.18). The SL > SS principle became a standard in conservation planning worldwide following its reiteration in the IUCN's (1980) highly influential World Conservation Strategy.

While there is more edge habitat there is also less core area in the several small scenario than in the single large scenario. Some species are core specialists (e.g. bonobo (*Pan paniscus*) and gibbon (Hylobatidae)) and, hence, would thrive better in the single large option. However, if conservation effort is focussed on species that are edge specialists (e.g. vervet monkey (*Chlorocebus pygerythrus*) which is found in savanna and woodland mosaics and olive baboon (*Papio anubis*) which is frequently seen at forest margins and open woodlands), the several small scenario is best.

There are three other assumptions in Figure 7.18 and the associated calculation of edge and core areas. Firstly, that all the habitat patches have straight edges and are square or rectangular, secondly that the edge is the same thickness at all points around the core area of an individual habitat patch, and thirdly that the species composition of each habitat patch is identical. In a real landscape it is unlikely that these assumptions will be valid. Nonetheless, the illustration is useful to make the point that there is no one solution that fits all conservation outcomes.

Simberloff and Abele (1976) considered that the theory of Island Biogeography is, in fact, agnostic on the SLOSS question. Other authors pointed out that the number of species in the several small or one or a few large patches will depend on the degree to which species composition varies among the small patches (i.e. the beta diversity – see Chapter 5) (Higgs and Usher 1980; Rösch et al. 2015). Fahrig (2020) reviewed the available literature from the past three decades and found that SS > SL in most cases. It is worth noting that most studies reviewed were of highly degraded and fragmented landscapes (such as that in northern Europe) in which there is very little remaining primary habitat (and, hence, few of the species that relied on it). The implication being that the SS > SL result typically reflects the preferences of the remaining generalists/edge specialists, as those species that would benefit from SL are absent.

Theoretical work related to the SLOSS question suggests a more complex picture, hence the switch to naming this issue the fragmenting-biodiversity debate. The newer name also broadens the issue to include ecological function as well as patch size and emphasises that ecosystems can be more adaptive and resilient when connected. This debate becomes important when making decisions about the conservation of species which are generally based on a combination of local knowledge and general rules or principles (reviewed in Gagné et al. 2015; Arroyo-Rodríguez et al. 2020).

7.4 CASE STUDY – TACKLING HABITAT LOSS

Human activities, including exploitation of natural resources, growing food, and building infrastructure and housing have left their mark across the globe. Associated with these human activities is habitat loss, fragmentation, and degradation impacts (Table 7.4).

The effects of habitat loss, fragmentation, and degradation are inherently scale and species dependent as well as context specific (Banks-Leite et al. 2020). In 2020, Riggio et al. published maps

Table 7.4 Definitions of habitat loss, fragmentation, and degradation

Habitat change	*Description*
Loss	The change of one habitat to another typically forest to agriculture and agriculture to urbanisation. In this case the forest habitat is lost. Habitat loss is typically the result of activities such as deforestation resulting from human activities or processes such as climate change.
Fragmentation	The breaking apart of a larger area of habitat which results in the same area of habitat being present as a number of smaller habitats patches, loss of connectivity, and increased edge effects.
Degradation	Caused by processes such as burning or selective logging and the gradual changes that occur in ecological communities over time due to unequal disruptions to the birth and death rates of different species, involves a reduction in the quality of a habitat without any significant changes to its amount or fragmentation.

of the Global human influence. Their analysis indicated that the human influence on the terrestrial surface of the globe was low on 48–56 per cent and very low on 20–34 per cent of the terrestrial surface of the globe. Much of the very low and low influence portions of the planet is comprised of cold (e.g. boreal forests, montane grasslands, and tundra) or arid (e.g. deserts) landscapes. In contrast, less than 1 per cent of temperate grasslands, tropical coniferous forests, tropical dry forests, tropical grasslands, mangroves, and montane grasslands have very low human influence across most or all the datasets they used in their analysis.

Habitat transformation (e.g. conversion to agriculture – Figure 7.19), overexploitation (e.g. hunting and overfishing), biotic exchange (e.g. invasive species – see Chapter 5), nutrient loading (e.g. pollution and increases in nitrogen – see Chapter 2), and climate change (see Chapter 2) are the main drivers of biodiversity loss. Not all ecosystems are affected equally. For example, tropical forests and grasslands or savannas (e.g. the Cerrado in Brazil) are being converted to agricultural land at a faster rate than any other ecosystem (Banks-Leite et al. 2020).

When addressing habitat loss and the associated ecological issues, it is necessary to draw on understanding developed across all

Figure 7.19 Habitat transformation in Sa Pa, Vietnam with rice cultivated in terraces replacing forest vegetation.

Source: photograph by Gina Hazell

chapters in this book. Knowledge of the habitat, the abiotic factors (Chapter 2), intra- and inter-specific behaviours (Chapters 3, 4, and 5), population structure and dynamics (Chapter 4), community composition (Chapter 5), and ecosystem functions (Chapter 6) sits alongside an understanding of the social systems involved (Chapter 6) and the routes towards policy formation and delivery (Chapter 1).

There are two basic options for restoring a particular habitat patch in a landscape: extending or repairing. Extending a habitat primarily helps counteract the effects of habitat loss and fragmentation, while repairing a habitat helps to counteract the effects of habitat degradation (Banks-Leite et al. 2020). Extending a habitat involves increasing the size of that habitat or adding new habitat patches near to extant patches of the same habitat to improve landscape-scale connectivity. Repairing a habitat involves techniques such as establishing buffer zones and habitat corridors, reducing browsing, species removal, and enrichment planting.

Restoration, according to the current, standard definition, is the practice of assisting the recovery of an ecosystem that has been degraded, damaged, or destroyed (Gann et al. 2019). This is a very wide definition, which means the concept of rewilding – commonly understood to be facilitating the development of a self-sustaining, self-organising, and resilient ecosystem shaped by natural processes (Pettorelli and Bullock 2023) – may be regarded as one approach to restoration.

However, other definitions of both restoration and rewilding exist. Restoration is commonly understood and implemented as the active process of assisting the recovery of an ecosystem that has been degraded, damaged, or destroyed back to its original state (Pettorelli and Bullock 2025). Restoration is, thus, expected to have a focus on recreating and maintaining the species community that would have been present had the ecosystem not been compromised (du Toit and Pettorelli 2019). This involves site preparation, the introduction of species, and long-term management to maintain benchmark conditions and, as a result, is costly and consequentially it is implemented at relatively small scales (Bullock et al. 2022). It is predicted that restoration of the planet's most degraded areas in combination with the protection of biodiversity hotspots could significantly boost carbon sequestration capacity and prevent about 70 per cent of predicted human-induced species extinctions (Strassburg et al. 2020).

The idea of rewilding first emerged in the 1980s but only really took off in the second half of the 2010s (Pettorelli et al. 2018) (Figure 7.20). This was at a time when concepts of ecosystem services and social-ecological systems (see Chapter 6) had gained traction and were reshaping the conceptualisation of nature and nature governance. Rewilding has many definitions but central to the concept is a reliance on natural processes, rather than management, to achieve desired outcomes. This can include the active reintroduction of species or a more passive approach where species are allowed to recolonise an area without human assistance (i.e. passive rewilding) (Lorimer et al. 2015; Falcón and Hansen 2018; du Toit and Pettorelli 2019; Carver et al. 2021) (Table 7.5).

Restoration and rewilding both have a role to play in addressing issues of biodiversity loss and climate change. It is not a case of adopting one approach or the other but adopting both depending on the situation.

Trophic rewilding is an approach that looks at the top-down regulation of lower trophic levels (see Chapters 5 and 6). This can be by introducing carnivores to reduce the number of herbivores or

Figure 7.20 An example of small-scale rewilding of a once overgrazed monoculture dairy pasture that is now Wild Finca, Asturias, Spain (www.wildfinca.com).

Source: photograph by Luke Blazejewski

Table 7.5 Comparison of the main characteristics of traditional restoration and rewilding

Characteristics	*Restoration*	*Rewilding*
Focus	Creation of a target ecological community	Resilience and adaptation
Aims	Safeguarding specific species and habitats	Facilitation of the emergence of self-sustaining, self-organising and resilient ecosystems shaped by natural processes.
Reliance on historical benchmarks	Higher	Lower
Outcomes	Predictable	Not very predictable
Management	Tends to be continuous	Little or none
Costs per hectare	High	Low
Area targeted	Small areas	Larger areas

Source: based on Pettorelli and Bullock (2025)

introducing herbivores to reduce the amount of vegetation (e.g. the re-introduction of grey wolf (*Canis lupus*) to Yellowstone National Park, MT, USA). Many nature-based solutions (see Chapter 6) that aim to mitigate climate change are based on the premise that certain taxa have consequential effects on ecosystem carbon capture and storage. Hence, there is a focus on protecting and restoring plants (primarily trees, mangroves, and seagrasses), and soil and sediment microbes. However, medium-to-large-bodied mammals also have a diversity of functional roles in the ecosystem, some of which can affect carbon sequestration and storage. For example, some of these animals are involved in the dispersal and germination of seeds from large-seeded trees (such trees frequently have carbon-dense wood) and herbivores can reduce plant competition and enhance soil nutrient supply and organic carbon storage (Schmitz et al. 2023).

Context is important when considering restoration programmes. There are differences between Europe, North America, and other parts of the world. The European landscape is dominated by agriculture and human related infrastructure such as canals, roads, and settlements. There are few large herbivores and carnivores and where these are present, their distribution is limited. In a European context, therefore, rewilding implies wildlife restoration in abandoned agricultural lands. In North America, despite the spread of agriculture and urbanisation, there are large areas of natural habitat in which few people live, and which contain large herbivores and carnivores. In this context, rewilding involves re-establishing wildlife populations in remnant natural habitats. In countries such as India, people live side-by-side with large herbivores and carnivores. Two examples illustrate this point. The Asian elephant (*Elephas maximus*) uses long established trails to move from one area to another. These routes frequently pass close to or through areas where people live and farm. Secondly, there is a significant population of leopards (*Panthera pardus*) living in and around Mumbai. The leopards spend the day outside the city and venture in at night to hunt. Their prey includes domestic dogs, cats, and pigs which are valued by people. In India, and in similar contexts across Asia and Africa, range expansion of wild animals into their former range(s) through dispersal is an implicit form of passive rewilding. In such instances it is the local communities who bear the costs of these human-wildlife conflicts. It is important, therefore, that the social-ecological system (see Chapter 6) is examined fully and ecological work is undertaken 'with' and preferably 'by' those local communities, appropriate policies and actions are put in place (see Chapter 1).

One conservation strategy that is currently garnering attention is reintroductions. Species reintroductions can be a tool for nature recovery through a compositional and/or functionalist approach. Compositionally, reintroductions restore trophic levels (see Chapter 5) through restoring communities. Functionally, reintroductions reinstate natural processes and their trophic cascades (see Chapters 5 and 6). This is achieved either by driving top-down trophic cascades or by reestablishing factors such as a landscape of fear (a concept that describes how the perception of predation risk influences the spatial and temporal behaviour of prey species in their habitats) and predation (e.g. reintroduction of the grey wolf (*Canis lupus*) and Eurasian lynx (*Lynx lynx*) in countries in Europe where they have become extinct), or through a bottom-up approach of engineering ecosystems (e.g. the reintroduction of European beaver (*Castor fiber*) into European wetland systems). Such a functionalist approach often necessitates a landscape-scale approach and a willingness to embrace open-ended outcomes.

A review was undertaken to explore barriers and opportunities to large-scale nature restoration and rewilding projects and help identify what works and issues encountered (Underwood et al. 2021) (Box 7.1). Writing in an African context, but highlighting issues relevant across many parts of the globe, Bezeng et al. (2025) proposed five recommendations for enhancing biodiversity conservation: (i) accelerate data collection, data sharing and analytics for informed policy and decision-making (see Chapter 1); (ii) innovate education and capacity building for future generations; (iii) enhance and expand protected areas, ecological networks and foundational legal frameworks (see this Chapter); (iv) unlock creative funding channels for cutting-edge conservation initiatives; and (v) integrate indigenous and local knowledge into forward-thinking conservation strategies.

In many ways, these barriers and opportunities bring this book back, full circle, to the opening chapter in which the importance of good governance was emphasised. Ecologists work under licence from local, national, and international communities. An ecologist delivers knowledge about a species, population, community, ecosystem, or landscape that is valued by the wider community.

BOX 7.1 BARRIERS AND OPPORTUNITIES TO LARGE-SCALE NATURE RESTORATION AND REWILDING PROJECTS (MODIFIED FROM UNDERWOOD ET AL. 2021)

Key findings

1. There is a range of motivations for large-scale nature restoration including those related to the interests of particular landowners with a rewilding vision, to responses to the challenges of working difficult sites, or to the delivery of ecosystem services such as water purification and flood management for ecosystem service delivery.
2. While some examples of rewilding are on owner-occupied sites where there is direct control over all aspects of the project other examples involve partnerships of many organisations. While decision making can be slowed down in large partnerships, projects can be carried out at a larger scale, gain more influence, and incorporate the views of all stakeholders.
3. Working with natural processes requires an open and flexible approach, as ecosystems are dynamic and constantly changing.
4. Views found within some of the large partnership projects include contrasting conservation interests which need to be reconciled.
5. Engagement with local communities, land managers, and other stakeholders is critical and should be maintained.
6. It is important to learn from other initiatives.
7. Empowering existing actors is important because they are generally embedded in the local landowning community and understand a local area well.
8. Economic benefits from rewilding include tourism, developing and marketing high quality, local products as a means of enhancing agricultural income, reduction in flood risk, increased protection of properties, and reduced maintenance costs of buildings and other infrastructure.
9. Changes in employment associated with management of the land and the wider economy include the creation of a range of new positions in land and habitat management, visitor management, education, and diversified businesses.

10. Funding is often a barrier to the scale and legacy of a restoration project.
11. Rewilding should work alongside other interventions and land management practices (e.g. on enclosed farmland land may be allocated for rewilding on poorer soils, alongside traditional conservation management, and productive agriculture using regenerative practices on the better land).
12. Large-scale nature restoration and rewilding projects benefit people locally and in wider society, using nature restoration to support improved quality of life.

This knowledge leads to a better societal understanding of how species are related to each other, how they behave, and how they are surviving and adapting to the ever-changing environment. Ecological knowledge and understanding has a valued role in improving the lives of those who live on the earth and those who will follow.

7.5 SELF-STUDY SUGGESTIONS

7.5.1 SELF-STUDY QUESTIONS

1. What is landscape ecology?
2. Outline the island biogeography theory, and the role played by island size and distance to a source of colonists.
3. What is metapopulation theory?
4. What is the patch–corridor–matrix model?
5. What is the rescue effect?
6. What is the difference between an edge and an ecotone?
7. What is the meaning of the term 'edge effect', and why is it important in ecology?
8. What do these words mean habitat deterioration, fragmentation, and loss, and why are they important?
9. What are the major forces driving habitat fragmentation?
10. How do habitat patches in a terrestrial habitat mosaic environment differ from island in an ocean?
11. How are patch size, interior species, and species diversity related?
12. What are the differences between a population, a community, an ecosystem, and a landscape.

7.5.2 A DEEPER DIVE

A. Search the internet of a list of species introductions. Which species are being reintroduced in the continent where you live and/or study?
B. Survey the size and shape of patches in your local landscape.

7.6 REFERENCES

Arroyo-Rodríguez, V., et al. (2020). 'Designing optimal human-modified landscapes for forest biodiversity conservation'. *Ecology Letters* 23: 1404–1420. https://doi.org/10.1111/ele.13535.

Banks-Leite, C., Ewers, R.M., Folkard-Tapp, H. and Fraser, A. (2020). 'Countering the effects of habitat loss, fragmentation, and degradation through habitat restoration'. *One Earth* 3(6): 672–676. https://doi.org/10.1016/j.oneear.2020.11.016.

Bezeng, B.S., et al. (2025). 'An African perspective to biodiversity conservation in the twenty-first century'. *Philosophical Transactions of the Royal Society B* 380: 20230443. http://doi.org/10.1098/rstb.2023.0443.

Blake, J.G. and Karr, J.R. (1984). 'Species composition of bird communities and the conservation benefit of large versus small forests'. *Biological Conservation* 30(2): 173–187. https://doi.org/10.1016/0006-3207(84)90065-X.

Brayley, O. (2022). 'Illuminating the problem of light pollution'. *The Niche* 53(4): 21–25.

Brown, J.H. and Kodrich-Brown A. (1977). 'Turnover rates in insular biogeography: Effect of immigration on extinction'. *Ecology* 58: 445–449. https://doi.org/10.2307/1935620.

Bullock, J.M., Fuentes-Montemayor, E., McCarthy, B., Park, K., Hails, R.S., Woodcock, B.A., Watts, K., Corstanje, R. and Harris, J. (2022) 'Future restoration should enhance ecological complexity and emergent properties at multiple scales'. *Ecography* e05780. https://doi.org/10.1111/ecog.05780.

Carver, S., et al. (2021). 'Guiding principles for rewilding'. *Conservation Biology* 35(6): 1882–1893. https://doi.org/10.1111/cobi.13730.

Clark, W. (2010). 'Principles of landscape ecology'. *Nature Education Knowledge* 3(10): 34.

Darlington, P.J., Jr. (1957). *Zoogeography: The Geographical Distribution of Animals*. New York: Wiley.

Diamond, J.M. (1976). 'Island biogeography and conservation: strategy and limitations'. *Science* 193: 1027–1029. https://doi.org/10.1126/science.193.4257.1027.

du Toit, J.T. and Pettorelli, N. (2019). 'The differences between rewilding and restoring an ecologically degraded landscape'. *Journal of Applied Ecology* 56(11): 2467–2471.

Fahrig, L. (2020). 'Why do several small patches hold more species than few large patches?' *Global Ecology and Biogeography* 29: 615–628. https://doi.org/10.1111/geb.13059.

Falcón, W. and Hansen, D.M. (2018). 'Island rewilding with giant tortoises in an era of climate change'. *Philosophical Transactions of the Royal Society B* 373: 20170442.

Farina, A. (1998). *Principles and Methods in Landscape Ecology*. New York: Chapman and Hall.

Forman, R.T.T. (1981). Interactions among landscape elements: a core of landscape ecology. In *Perspectives in Landscape Ecology: Proceeding of the International Congress of Landscape Ecology 1981*. Wageningen, The Netherlands: Veldhoven.

Forman, R.T.T. (1986). *Landscape Ecology*. New York: John Wiley.

Forman, R.T.T. (1995). *Land Mosaics: The Ecology of Landscapes and Regions*. Cambridge: Cambridge University Press.

Forman, R.T.T. and Godron, M. (1981). 'Patches and structural components for a landscape ecology'. *BioScience* 31(10): 733–740. https://doi.org/10.2307/1308780.

Freemark, K.E. and Merriam, H.G. (1986). 'Importance of area and habitat heterogeneity to bird assemblages in temperate forest fragments'. *Biological Conservation* 36: 115–141.

Gagné, S.A., Eigenbrod, F., Bert, D., Cunnington, G.M., Olson, L.T., Smith, A.C. and Fahrig, L. (2015). 'A simple landscape design framework for biodiversity conservation'. *Landscape and Urban Planning* 136: 13–27. https://doi.org/10.1111/geb.13059.

Gann, G. D., et al. (2019). 'International principles and standards for the practice of ecological restoration (second edition)'. *Restoration Ecology* 27: S1–S46. https://doi.org/10.1111/rec.13035.

Graham, M. and Socha, J.J. (2018). 'Going the distance: the biomechanics of gap-crossing behaviours'. *Journal of Experimental Biology* 2019: 1–14. https://doi.org/10.1002/jez.2266.

Higgs, A.J. and Usher, M.B. (1980). 'Should nature reserves be large or small?' *Nature* 285: 568–569. https://doi.org/10.1038/285568a0.

IUCN. (1980). *World Conservation Strategy: Living Resource Conservation for Sustainable Development*. Gland, Switzerland: International Union for Conservation of Nature and Natural Resources.

Karr, J.R. (1982). 'Production and energy pathways, and community diversity in forest birds'. In F. Golley and E. Medina (eds), *Tropical Ecological Systems: Trends in Terrestrial and Aquatic Research*, 161–176. New York: Springer-Verlag.

Kritzer, J. P. and Sale, P. F. (eds). (2006). *Marine Metapopulations*. New York: Academic Press.

Levins, R. (1969). 'Some demographic and genetic consequences of environmental heterogeneity for biological control'. *Bulletin of the Entomological Society of America* 15(3): 237–240. https://doi.org/10.1093/besa/15.3.237.

Lorimer, J., Sandom, C., Jepson, P., Doughty, C., Barua, M. and Kirby, K.J. (2015) 'Rewilding: science, practice, and politics'. *Annual Review of Environment and Resources* 40: 39–62. https://doi.org/10.1146/annurev-environ-102014-021406.

MacArthur, R.H. and Wilson, E.O. (1963). 'An equilibrium theory of insular zoogeography'. *Evolution* 17: 373–387.

MacArthur, R.H. and Wilson, E.O. (1967). *The Theory of Island Biogeography*. Princeton, NJ: Princeton University Press

May, R.M. (1975). 'Island biogeography and the design of wildlife preserves'. *Nature* 254: 177–178. https://doi.org/10.1038/254177a0.

Murcia, C. (1995). 'Edge effects in fragmented forests: implications for conservation'. *Trends in Ecology and Evolution* 10: 58–62. 10.1016/S0169-5347(00)88977-6.

Patton, D.R. (1975). 'A diversity index for quantifying habitat "edge"'. *Wildlife Society Bulletin* 3: 171–173.

Pettorelli N., Barlow J., Stephens P.A., Durant S.M., Connor B., Schulte to Bühne H., Sandom C.J., Wentworth, J. and du Toit, J.T. (2018). 'Making rewilding fit for policy'. *Journal of Applied Ecology* 55(3): 1114–1125. https://doi.org/10.1111/1365-2664.13082.

Pettorelli, N. and Bullock, J.M. (2023). 'Restore or rewild? Implementing complementary approaches to bend the curve on biodiversity loss'. *Ecological Solutions and Evidence* 4: e12244. https://doi.org/10.1002/2688-8319.12244.

Pettorelli, N and Bullock, J.M. (2025). 'Rewilding and restoring: two distinct yet complementary approaches to nature recovery'. *British Wildlife* 36(6): 418–424.

Preston, F.W. (1962). 'The canonical distribution of commonness and rarity: parts 1 and 2'. *Ecology* 43: 185–215, 410–432.

Rigby, C. and Lawton, J.H. (1981). 'Species-area relationships of arthropods on host plants: herbivores on bracken'. *Journal of Biogeography* 8: 125–133. https://doi.org/10.2307/2844555.

Riggio, J., et al. (2020). 'Global human influence maps reveal clear opportunities in conserving Earth's remaining intact terrestrial ecosystems'. *Global Change Biology* 26: 4344–4356. https://doi.org/10.1111/gcb.15109.

Robbins, C.S., Dawson, D.K. and Dowell, B.A. (1989). *Habitat Area Requirements of Breeding Forest Birds of the Middle Atlantic States*. Wildlife Monographs No. 103. Hoboken, NJ: Wiley.

Rösch, V., Tscharntke, T., Scherber, C. and Batáry, P. (2015). 'Biodiversity conservation across taxa and landscapes requires many small as well as single large habitat fragments'. *Oecologia* 179: 209–222. https://doi.org/10.1007/s00442-015-3315-5.

Rosenberg, D.K., Noon, B.R. and Meslow, E.C. (1997). 'Biological corridors: form, function, and efficacy'. *Bioscience* 47(10): 677–687. https://doi.org/10.2307/1313208.

Schlaepfer, M.A., Runge, M.C. and Sherman, P.W. (2002). 'Ecological and evolutionary traps'. *Trends in Ecology & Evolution* 17(10): 474–480. https://doi.org/10.1016/S0169-5347(02)02580-6.

Schmitz, O.J., et al. (2023). 'Trophic rewilding can expand natural climate solutions'. *Nature Climate Change* 13: 324–333. https://doi.org/10.1038/s41558-023-01631-6.

Simberloff, D.S. and Abele, L.G. (1976). 'Island biogeography theory and conservation practice'. *Science* 191: 285–286. https://doi.org/10.1126/science.191.4224.285.

Sisk, T.D., Haddad, N.M. and Ehrlich, P.R. (1997). 'Bird assemblages in patchy woodlands: Modelling the effects of edge and matrix habitats'. *Ecology* 78: 1770–1780. https://doi.org/10.1890/1051-0761(1997)007[1170:BAIPWM]2.0.CO;2.

Strassburg, B. B. N., et al. (2020). 'Global priority areas for ecosystem restoration'. *Nature* 586: 724–729. https://doi.org/10.1038/s41586-020-2784-9.

Underwood, S, Kaczor, K, Roberts, V, Tooze, G, Rayment, M, Smith, M, Fry, C, Swindlehurst, S, Armstrong, S and James, N. (2021). *Mainstreaming Large Scale Nature Restoration*. Research Report No. 1271. Inverness: NatureScot.

Van Schmidt, N.D. and Beissinger, S.R. (2020). 'The rescue effect and inference from isolation–extinction relationships'. *Ecology Letters* 23(4): 598–606. https://doi.org/10.1111/ele.13460.

Weaver, M and Kellman, M. (1981). 'The effects of forest fragmentation on woodlot tree biotas in southern Ontario'. *Journal of Biogeography* 8: 199–210. https://doi.org/10.2307/2844677.

Whitcomb, R.F., Lynch, J.F., Opler, P.A. and Robbins, C.S. (1976) 'Island biogeography and conservation: strategy and limitations'. *Science* 193: 1030–1032. https://doi.org/10.1126/science.193.4257.1030.

With, K.A. (2019). 'An introduction to landscape ecology: foundations and core concepts'. In K.A. With, *Essentials of Landscape Ecology*, 1–13. Oxford: Oxford University Press. https://doi.org/10.1093/oso/9780198838388.003.0001.

INDEX

Page numbers in *italics* denote figures, those in **bold** denote tables.

For Product Safety Concerns and Information please contact our EU representative GPSR@taylorandfrancis.com
Taylor & Francis Verlag GmbH, Kaufingerstraße 24, 80331 München, Germany

www.ingramcontent.com/pod-product-compliance
Lightning Source LLC
LaVergne TN
LVHW012334130926
847902LV00028B/394
9781032994970